500++

MEDIUM-HARD

KAKURO

PUZZLE

BOOK

DJAP

500++ Medium-Hard Kakuro Puzzle Book

Puzzles by DJAPE

All puzzles in this book are new and not published elsewhere!

First edition: November 2022

ISBN 979-8-36292-982-4

Welcome to the book with more than 500 Kakuro puzzles!

Over the years, I have often received feedback from my readers that you like my Kakuro puzzles because they are quite challenging. Then, last year, I published one book only with the easy and medium puzzles. And now I've decided to go back, drop the easy, and publish a book only with the medium and hard Kakuros!

I will assume that you are already familiar with the Kakuro puzzles. If not, I strongly recommend that you first try that other book, titled "**1,000++ Easy-Medium Kakuro**". To find it, just Google this number: **9798769433528**.

Still, let me quickly remind you of the Kakuro rules: fill in the empty white squares with **numbers 1 to 9**, so that the sum of digits in each connected horizontal or vertical string of cells adds up to the number which is indicated in the corresponding black square above or to the left. To complicate things a little, there is one more rule: **a digit cannot be repeated within any sum**.

Kakuro puzzles come in different shapes and sizes. Regardless of the size, to solve Kakuros you need to analyze possible combinations of numbers for a given sum. For example:

- 7 in a string of 2 cells can be either 1+6, or 2+5, or 3+4;
- 8 over 3 cells can either be 1+2+5 or 1+3+4;
- 21 over 3 cells can be: 4+8+9 or 5+7+9 or 6+7+8.

Of course, you don't know immediately in which order the numbers will be placed in the correct solution. So, you must cross the information you get from one region with a perpendicular region and see if it helps you solve the intersecting number.

Example: a **4** over **2 cells** is intersecting a **6** over **2 cells**.

The 4 can be only **1+3** or **3+1**. If the number intersecting with the "6" sum is 1, then the other number in the "6" sum would be 5 (1+5=6). That seems ok.

However, if the intersecting number is 3, then the other number in the "6" sum would also have to be 3, because 3+3=6, but remember the rule which says that the repeats are not allowed! Therefore, the only possible solution for 4 (over 2 numbers) crossing the 6 (also over 2 numbers) is that the intersecting number is 1, and that the other two numbers are 3 (1+3=4) and 5 (1+5=6). Got it?

Other than analyzing the sums of numbers, there is a technique which can help a lot, yet many Kakuro solvers are unaware of it. It is called **"Splitting"**. This technique can be applied when a certain **RECTANGLE** (or **square**) is **completely covered** by a few horizontal or vertical sum regions.

Have a look at the upper left corner of this sample puzzle.

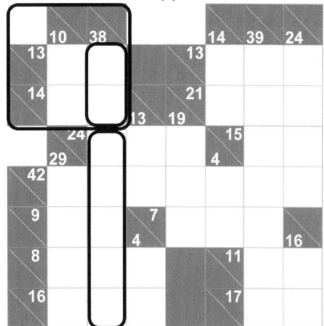

The highlighted 2x2 region is completely covered by two horizontal clues, the **"13"** and the **"14"**, which means that the sum of all 4 numbers of this 2x2 region is 14+13=27.

Now, we notice that the sum of the two numbers vertically is "10". So, we know the sum of four numbers (27), and we know the sum of two of those four numbers (10). We can conclude that the sum of the other two numbers is 27-10=17. Those two cells are circled.

So, now, we can **"split"** the vertical **"38"** sum into two: the top part consisting of two cells with the sum 17 and the bottom consisting of 5 cells with the sum 38-17=21. Got it? Since the sum of the two top numbers is 17, which can only be 9+8, the upper left corner is pretty much solved! (**hint**: the 10 cannot be 5+5)

This was a simple example; it can get much more complex than this, but now that you are aware of it, try searching for such regions and applying this technique. In fact, try and find one more region that can be split in this same puzzle.

One thing to always remember is: **DO NOT GUESS**! Look around, be patient, think, use deduction, arithmetic and logic and you will eventually prevail. If you guess you might go wrong. All puzzles in this book have one solution only and do not require guessing at all!

If you crave for more puzzle books, please visit my Amazon Brand Store at:

amazon.com/djape

Enjoy! 😊   **DJAPE**

#1 MEDIUM

#2 HARD

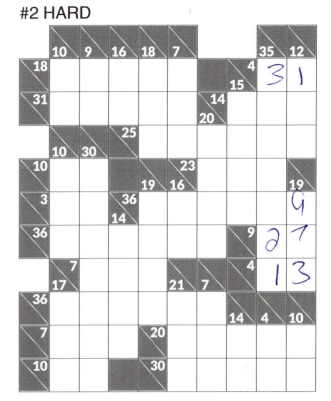

#3 MEDIUM

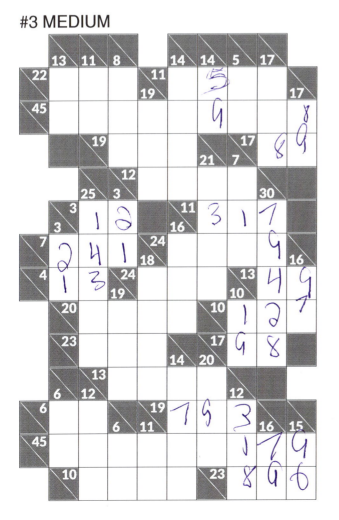

#4 HARD

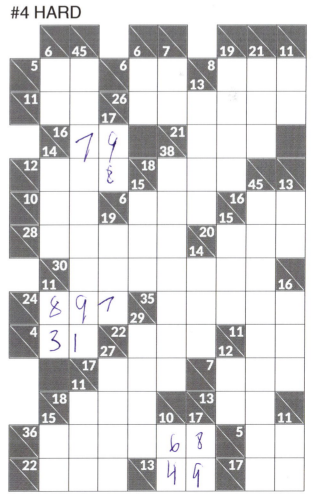

#5 MEDIUM

#6 HARD

#7 MEDIUM

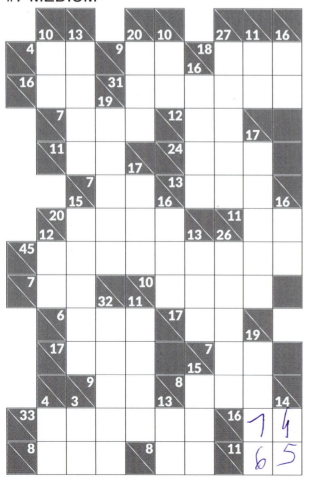

#8 HARD

#9 MEDIUM

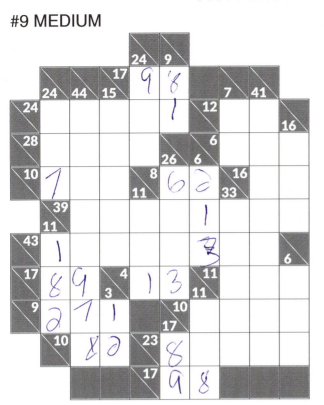

#10 HARD

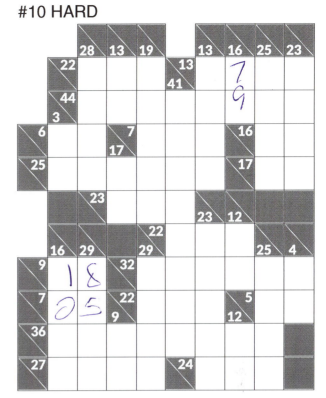

#11 MEDIUM

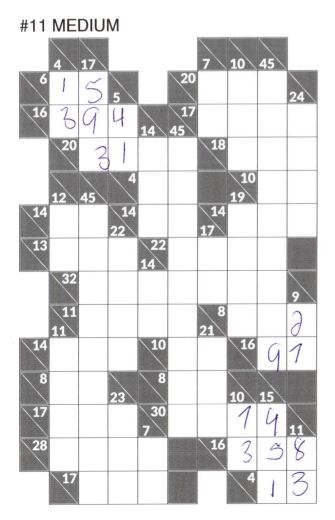

#12 HARD

#13 MEDIUM

#14 HARD

#15 MEDIUM

#16 HARD

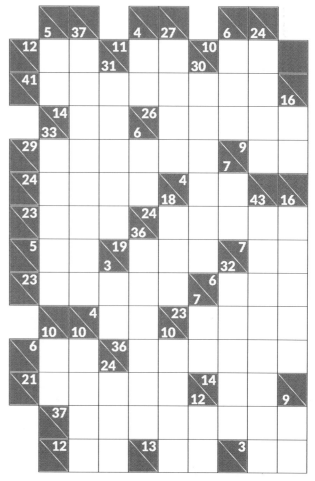

#17 MEDIUM

#18 HARD

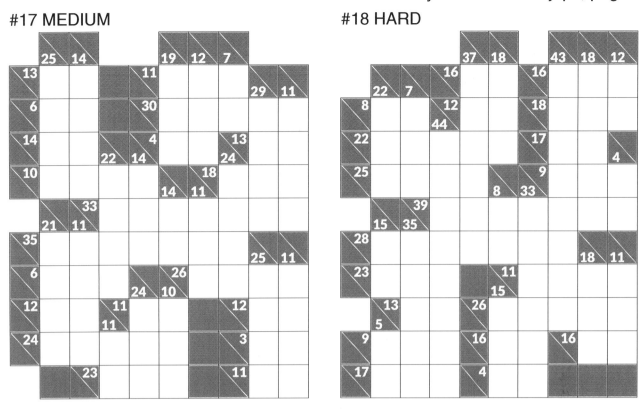

#19 MEDIUM

#20 HARD

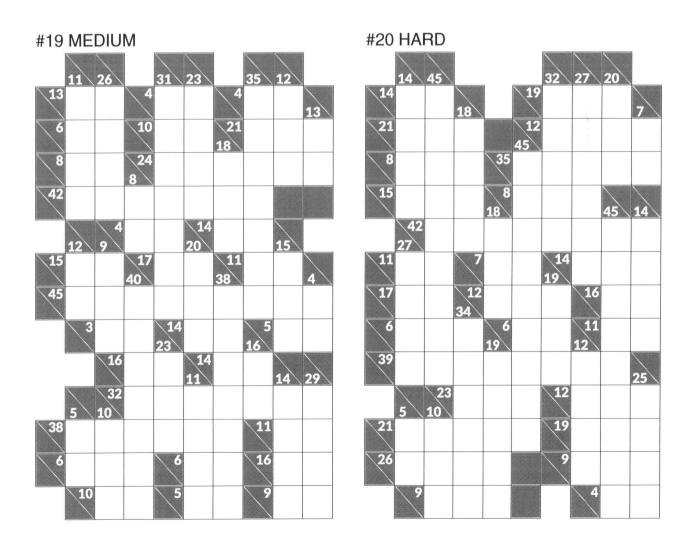

#21 MEDIUM

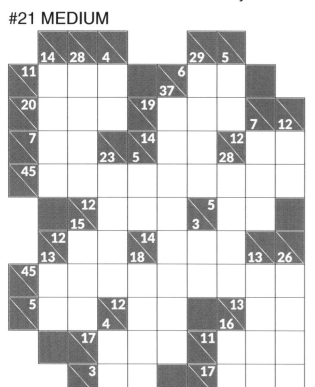

#22 HARD

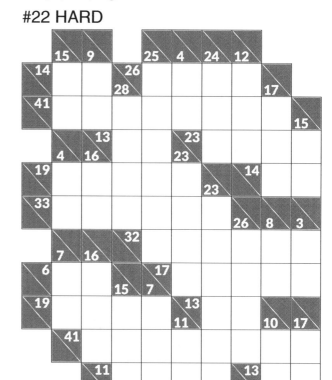

#23 MEDIUM

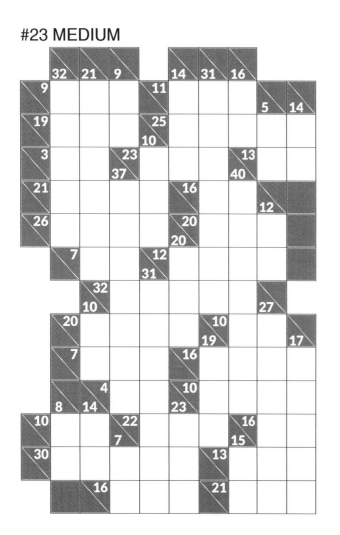

#24 HARD

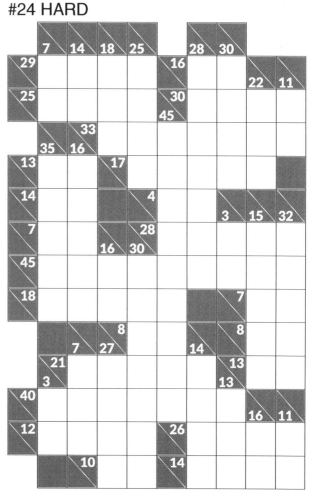

#25 MEDIUM

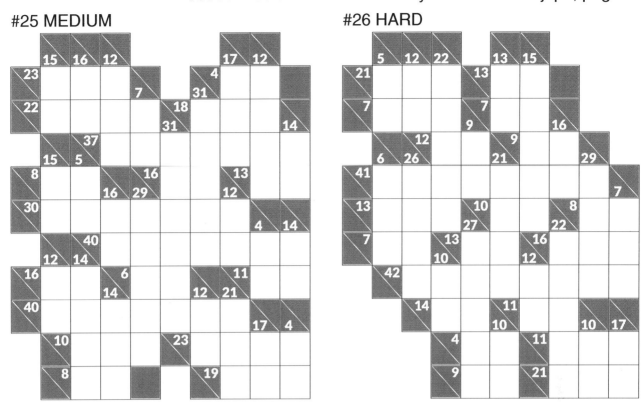

#26 HARD

#27 MEDIUM

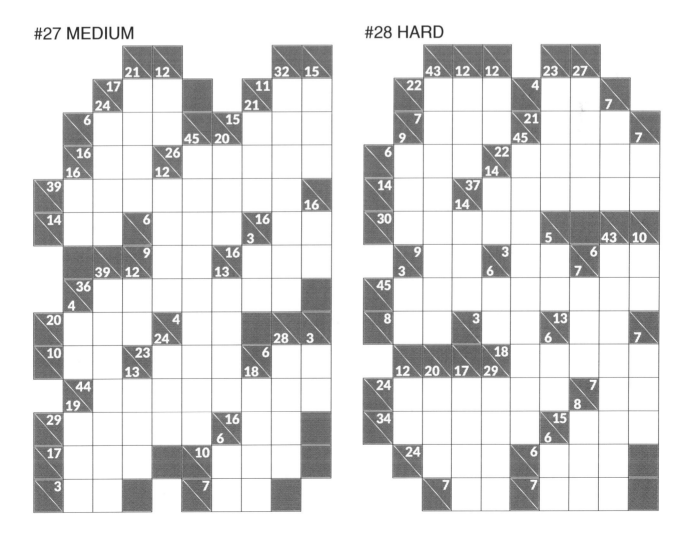

#28 HARD

#29 MEDIUM

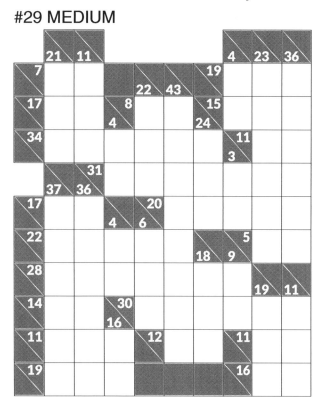

#30 HARD

#31 MEDIUM

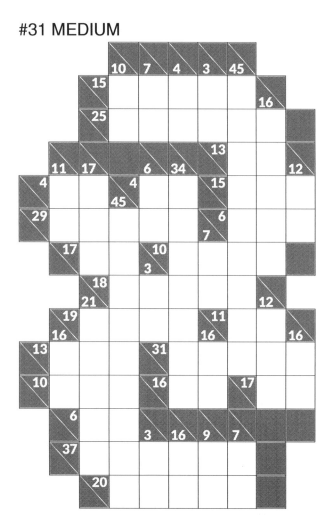

#32 HARD

#33 MEDIUM

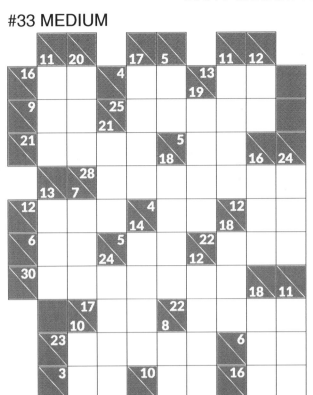

#34 HARD

#35 MEDIUM

#36 HARD

#37 MEDIUM

#38 HARD

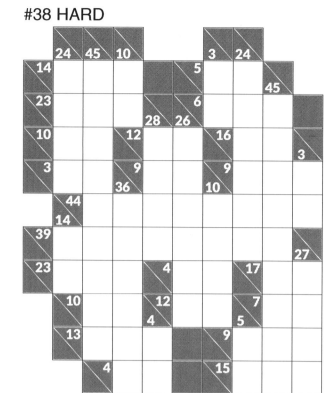

#39 MEDIUM

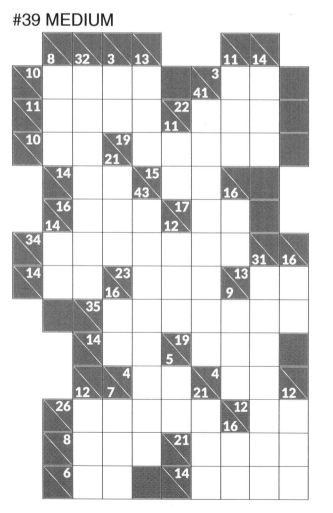

#40 HARD

#41 MEDIUM

#42 HARD

#43 MEDIUM

#44 HARD

#45 MEDIUM

#46 HARD

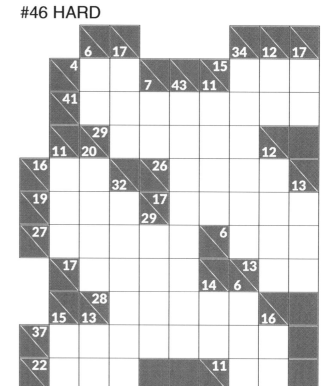

#47 MEDIUM

#48 HARD

#49 MEDIUM

#50 HARD

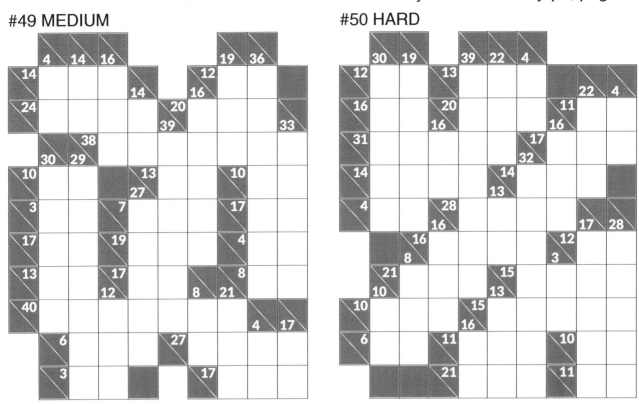

#51 MEDIUM

#52 HARD

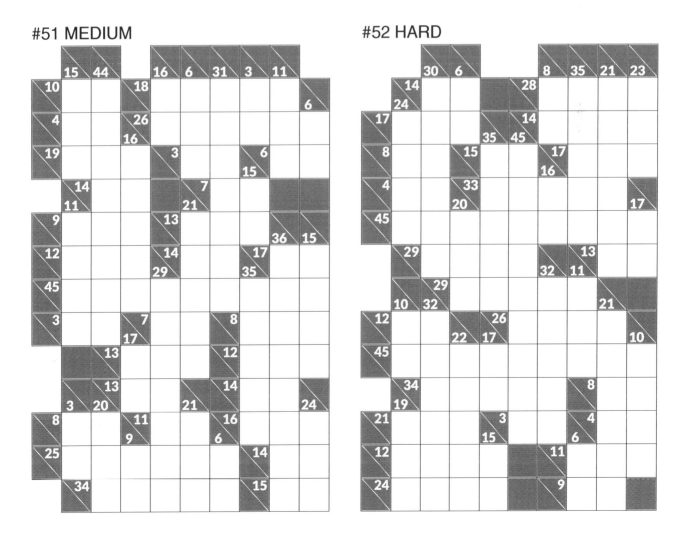

#53 MEDIUM

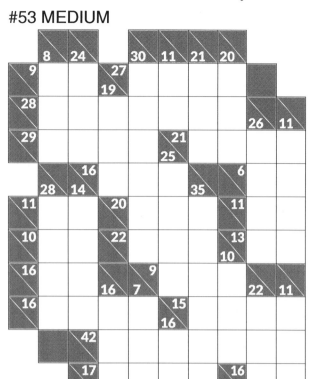

#54 HARD

#55 MEDIUM

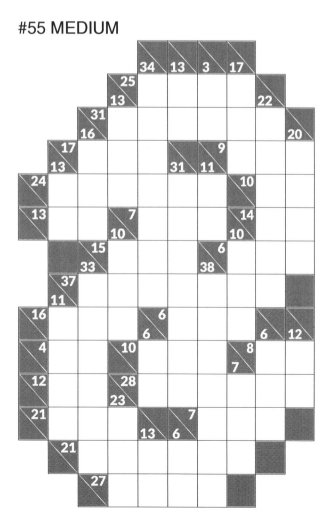

#56 HARD

#57 MEDIUM

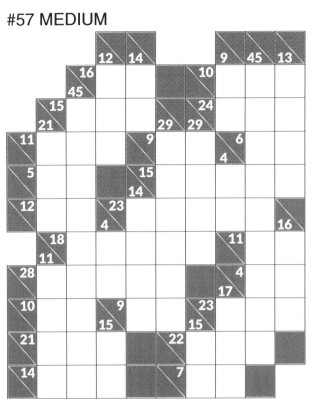

#58 HARD

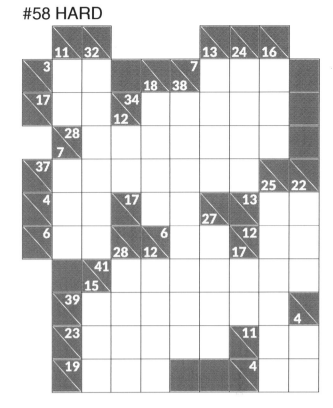

#59 MEDIUM

#60 HARD

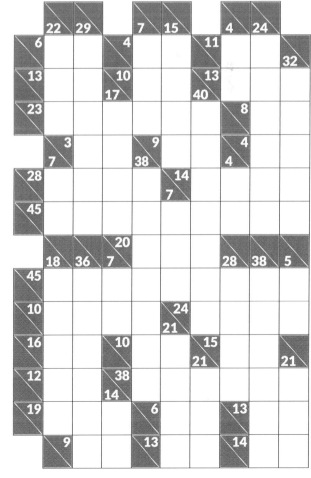

#61 MEDIUM

#62 HARD

#63 MEDIUM

#64 HARD

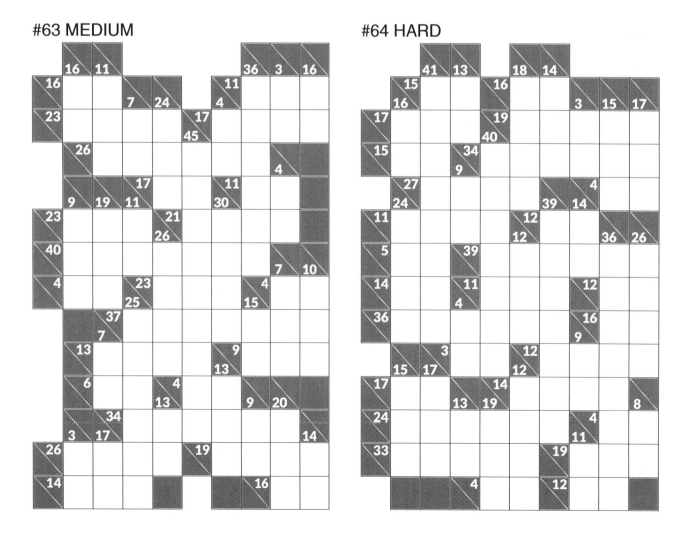

#65 MEDIUM

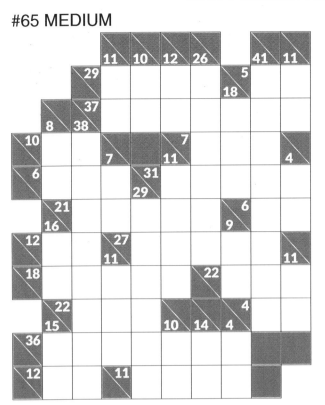

#66 HARD

#67 MEDIUM

#68 HARD

#69 MEDIUM

#70 HARD

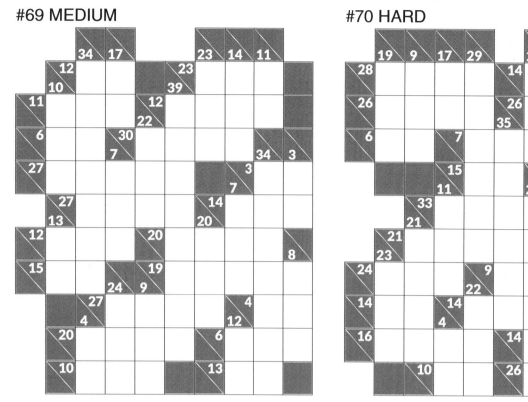

#71 MEDIUM

#72 HARD

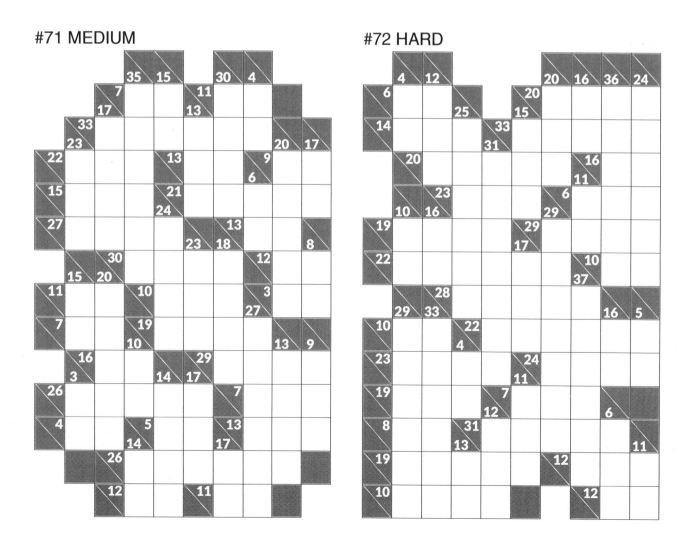

#73 MEDIUM

#74 HARD

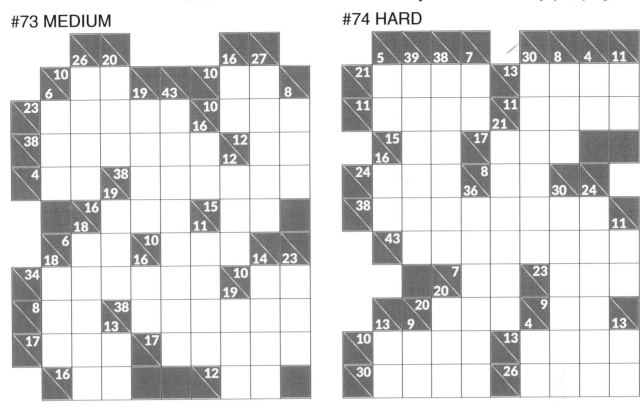

#75 MEDIUM

#76 HARD

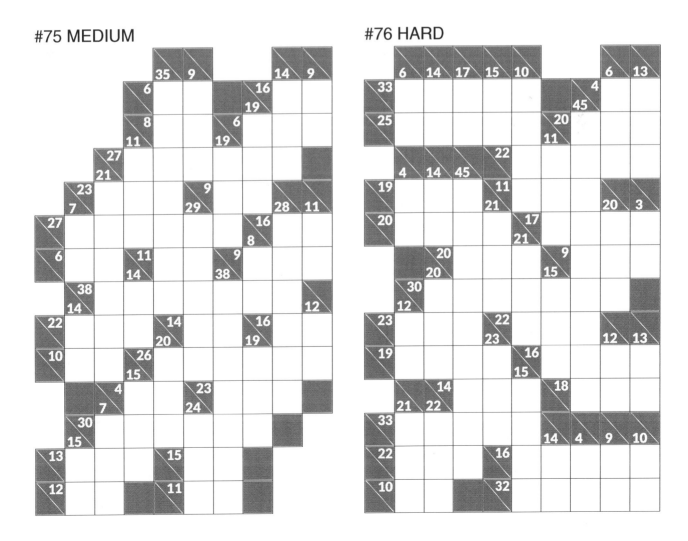

#77 MEDIUM

#78 HARD

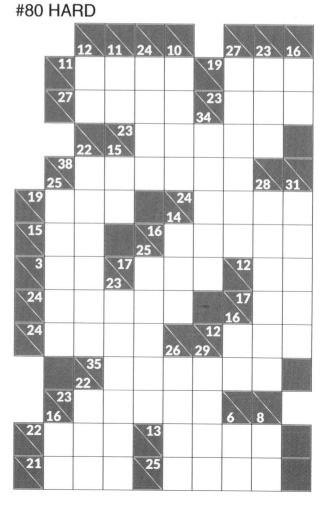

#79 MEDIUM

#80 HARD

#81 MEDIUM

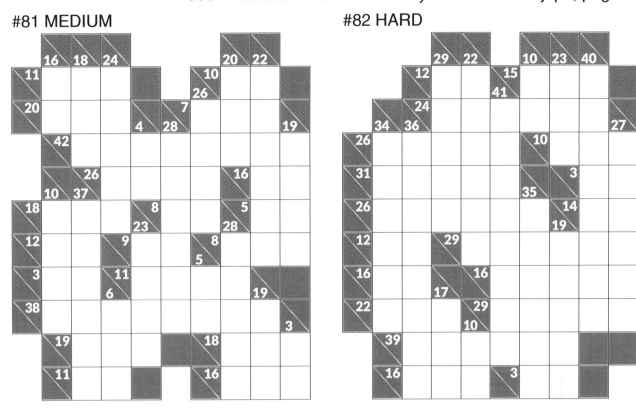

#82 HARD

#83 MEDIUM

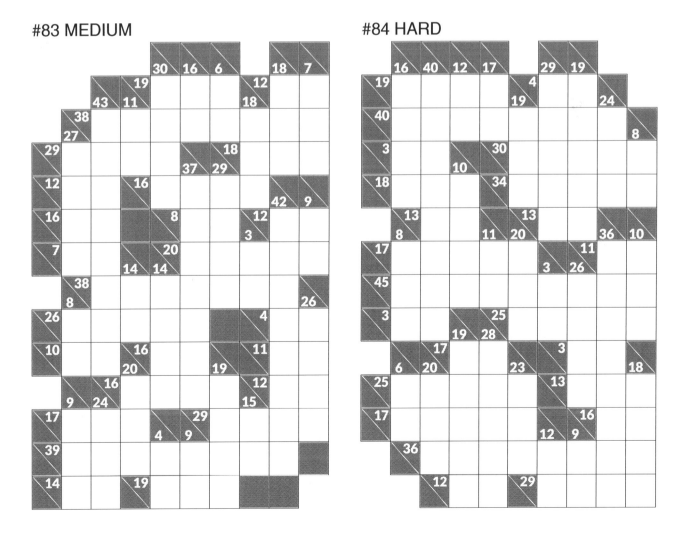

#84 HARD

#85 MEDIUM

#86 HARD

#87 MEDIUM

#88 HARD

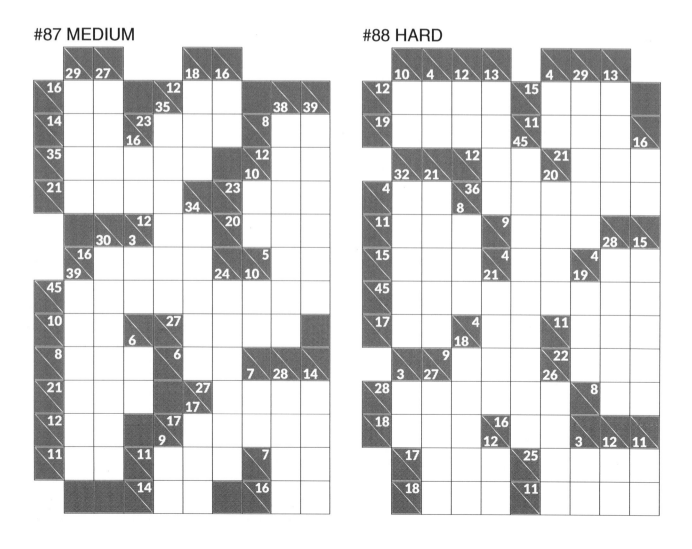

#89 MEDIUM

#90 HARD

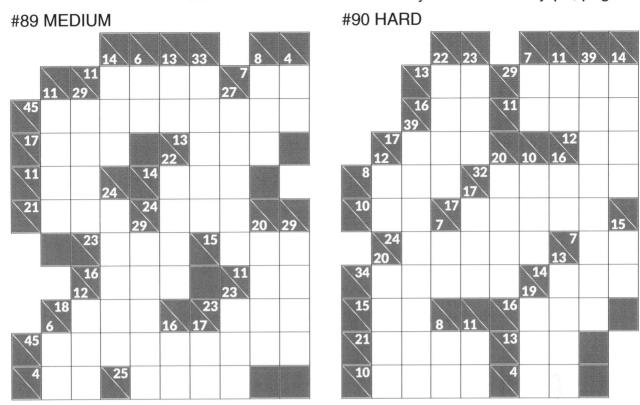

#91 MEDIUM

#92 HARD

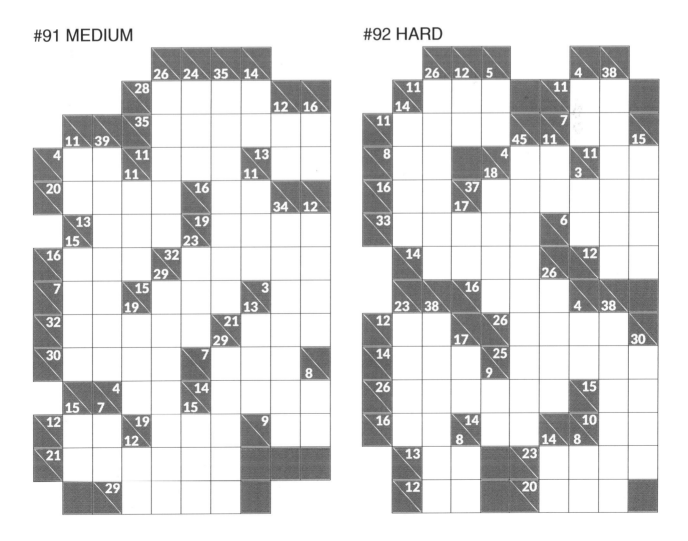

#93 MEDIUM

#94 HARD

#95 MEDIUM

#96 HARD

#97 MEDIUM

#98 HARD

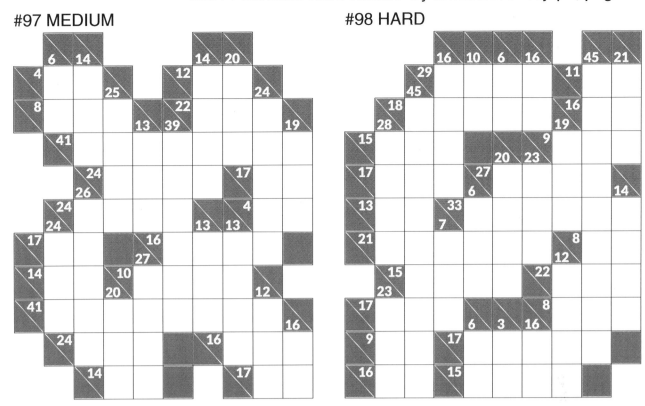

#99 MEDIUM

#100 HARD

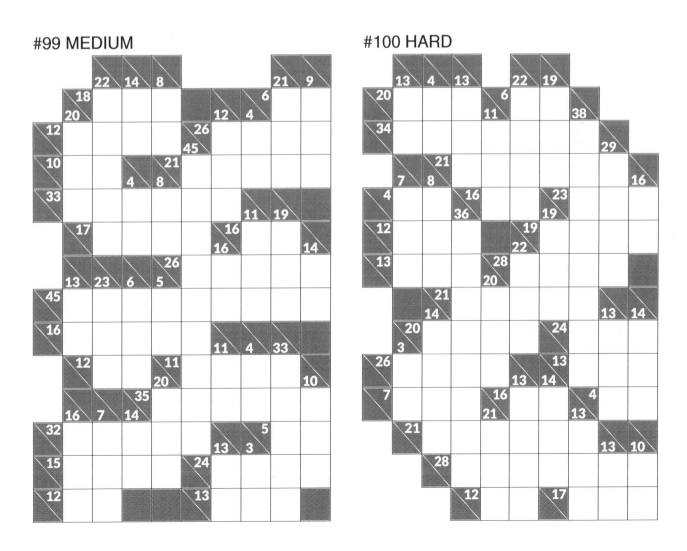

#101 MEDIUM

#102 HARD

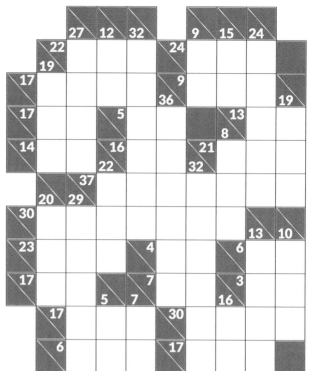

#103 MEDIUM

#104 HARD

#105 MEDIUM

#106 HARD

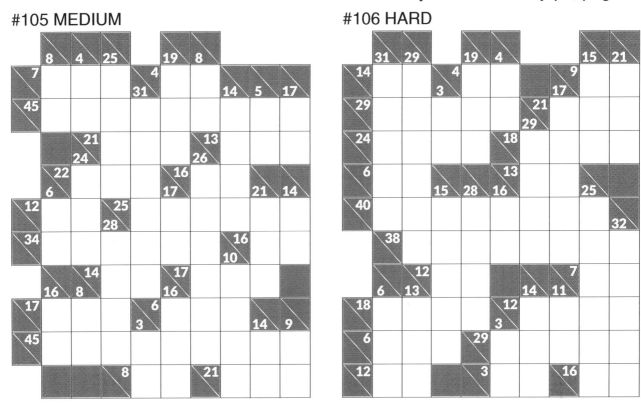

#107 MEDIUM

#108 HARD

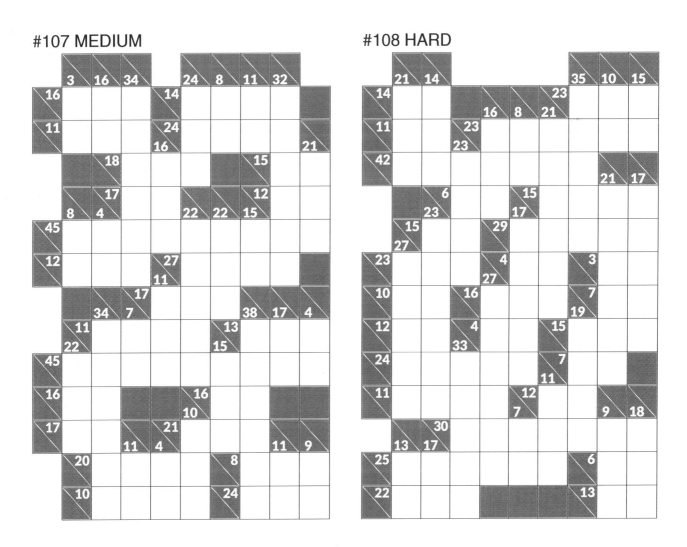

#109 MEDIUM

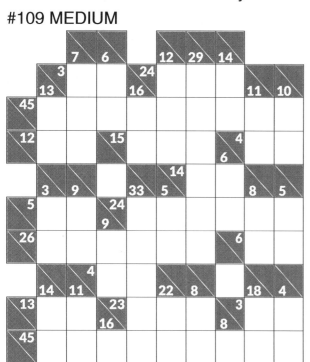

#110 HARD

#111 MEDIUM

#112 HARD

#113 MEDIUM

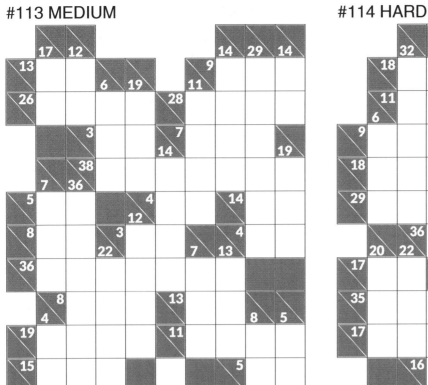

#114 HARD

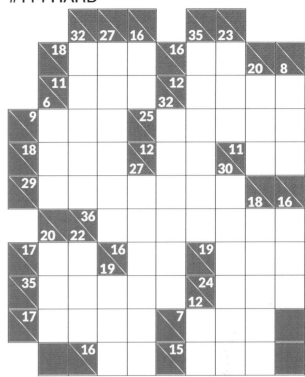

#115 MEDIUM

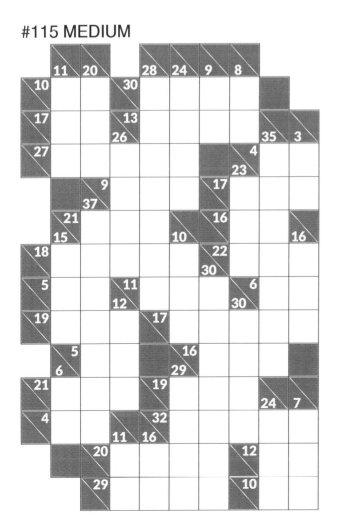

#116 HARD

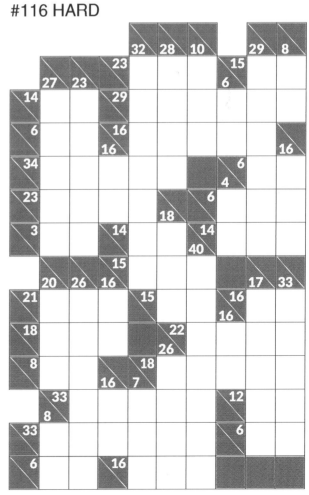

#117 MEDIUM

#118 HARD

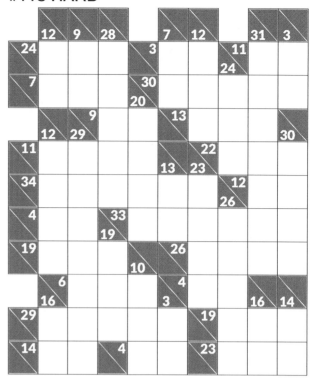

#119 MEDIUM

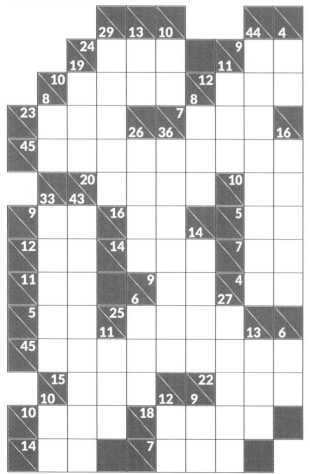

#120 HARD

#121 MEDIUM

#122 HARD

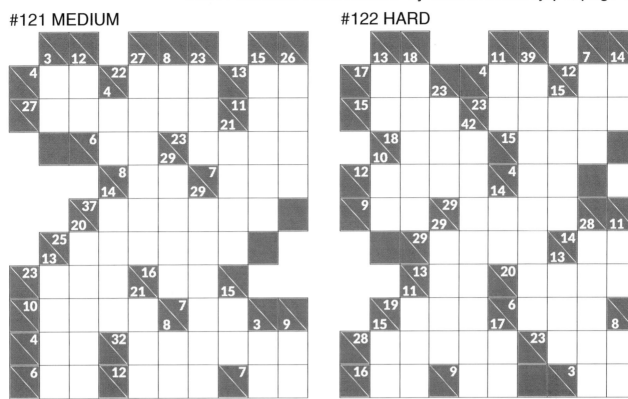

#123 MEDIUM

#124 HARD

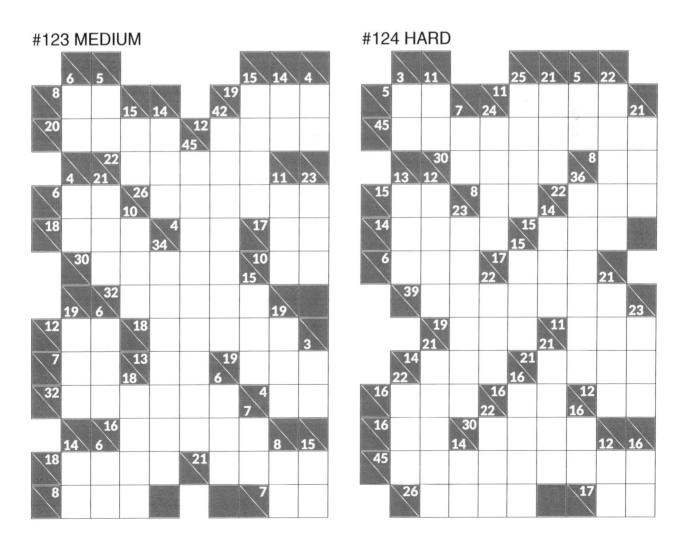

#125 MEDIUM

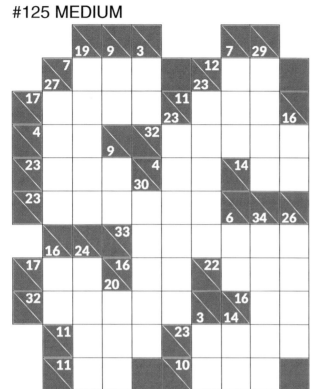

#126 HARD

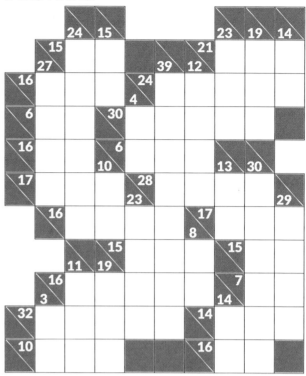

#127 MEDIUM

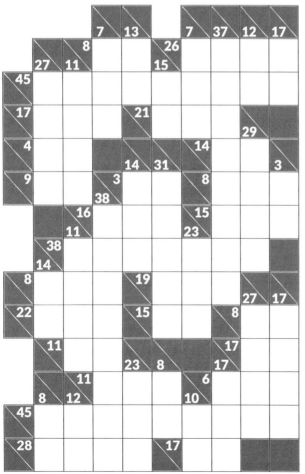

#128 HARD

#129 MEDIUM

#130 HARD

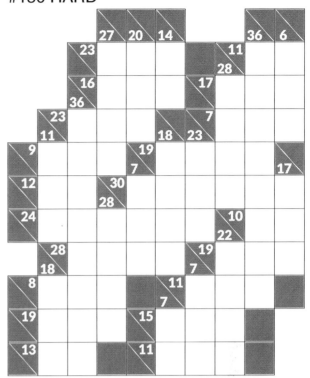

#131 MEDIUM

#132 HARD

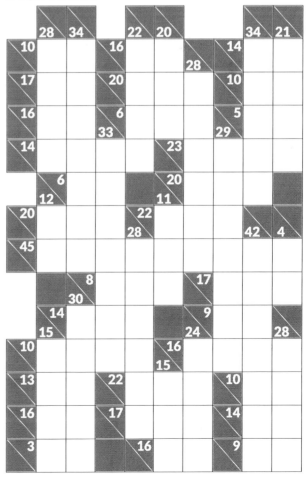

#133 MEDIUM

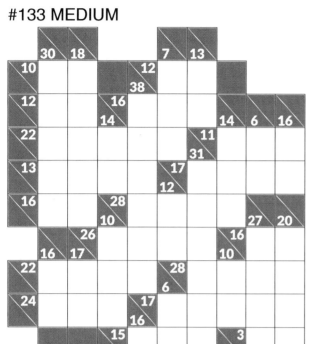

#134 HARD

#135 MEDIUM

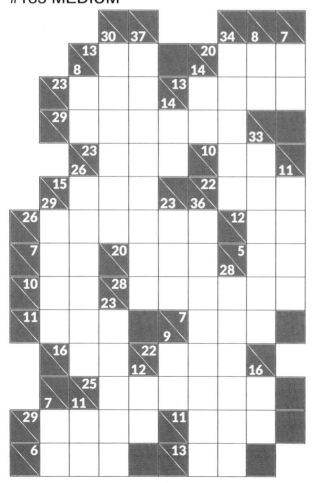

#136 HARD

#137 MEDIUM

#138 HARD

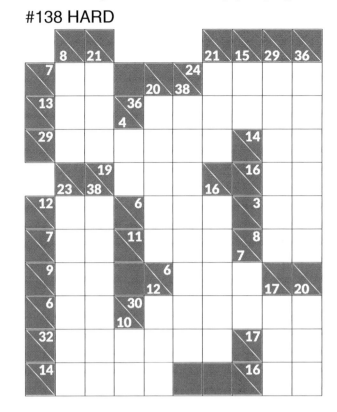

#139 MEDIUM

#140 HARD

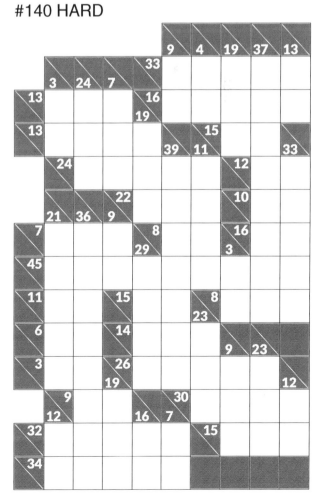

#141 MEDIUM

#142 HARD

#143 MEDIUM

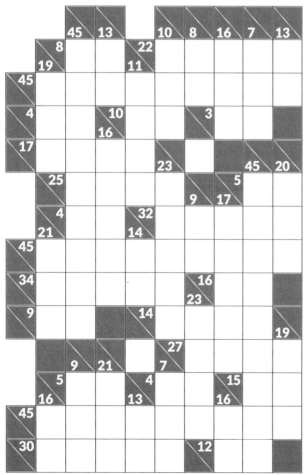

#144 HARD

#145 MEDIUM

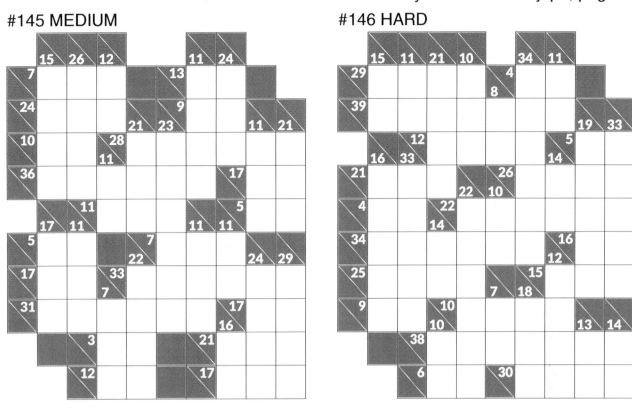

#146 HARD

#147 MEDIUM

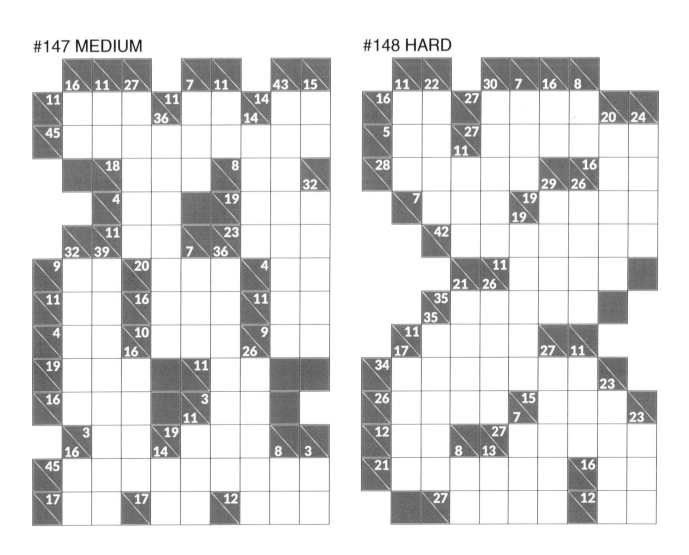

#148 HARD

#149 MEDIUM

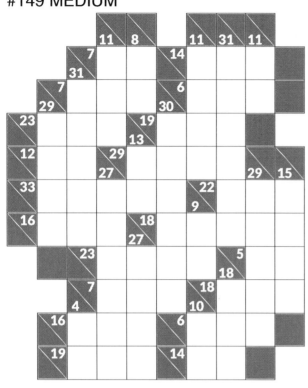

#150 HARD

#151 MEDIUM

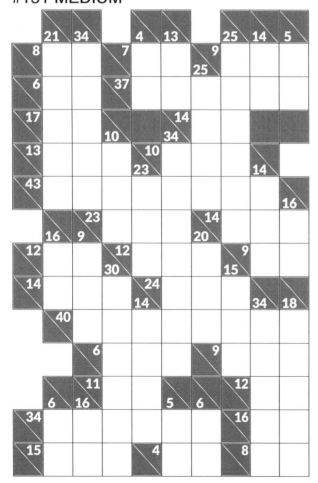

#152 HARD

#153 MEDIUM

#154 HARD

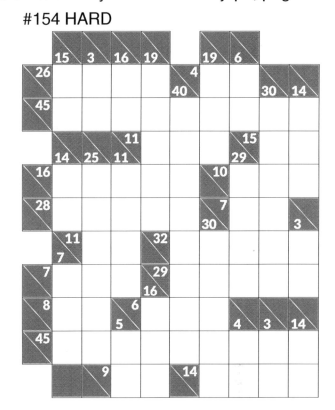

#155 MEDIUM

#156 HARD

#157 MEDIUM

#158 HARD

#159 MEDIUM

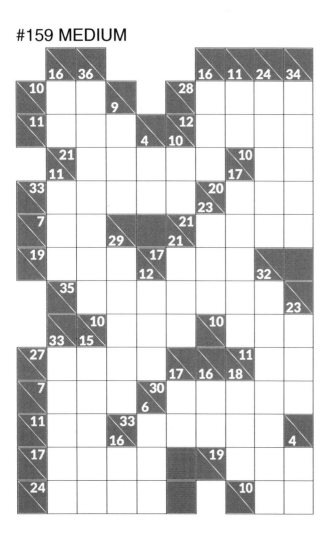

#160 HARD

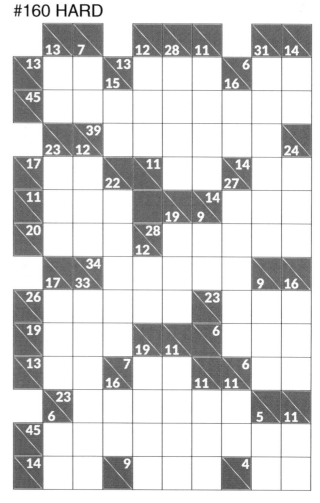

#161 MEDIUM

#162 HARD

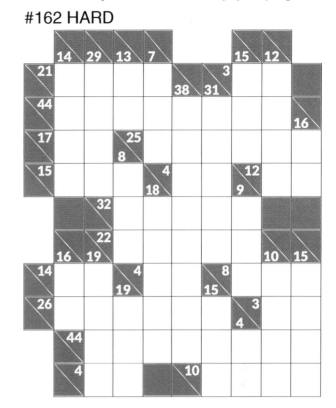

#163 MEDIUM

#164 HARD

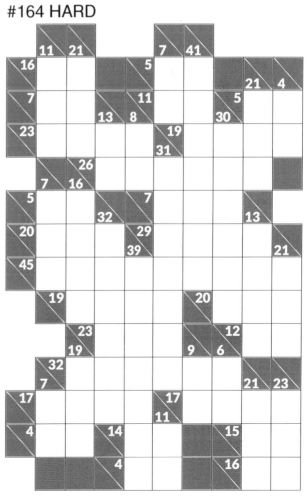

#165 MEDIUM

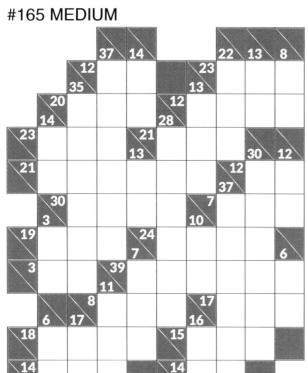

#166 HARD

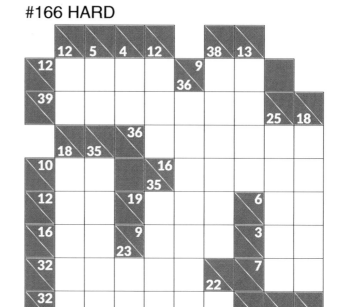

#167 MEDIUM

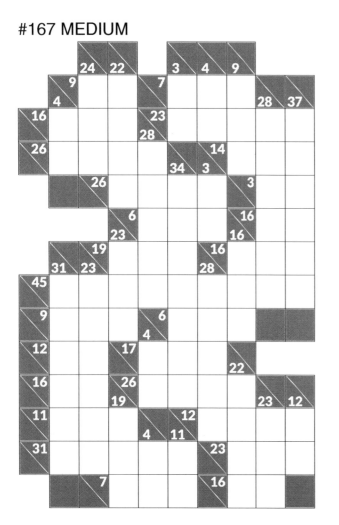

#168 HARD

#169 MEDIUM

#170 HARD

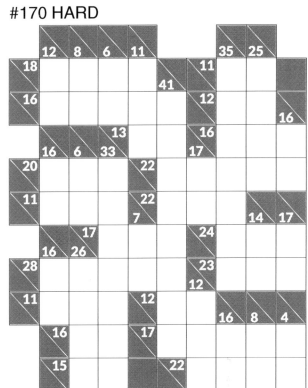

#171 MEDIUM

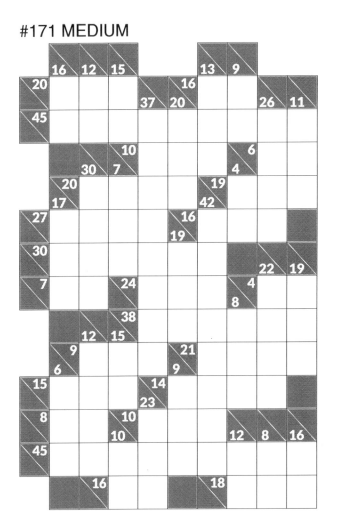

#172 HARD

#173 MEDIUM

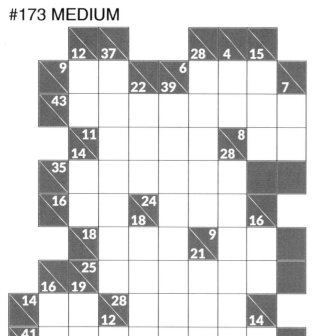

#174 HARD

#175 MEDIUM

#176 HARD

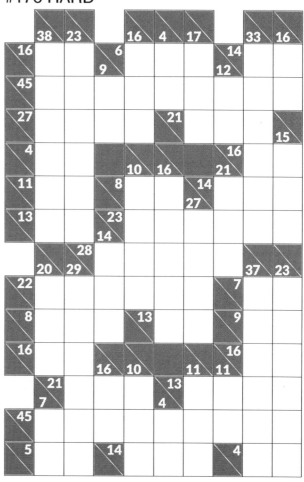

#177 MEDIUM

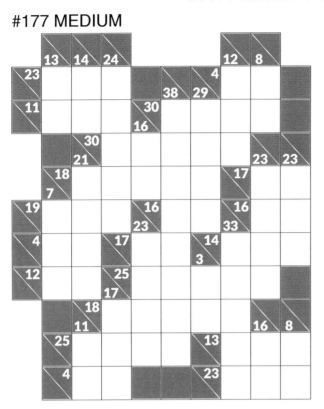

#178 HARD

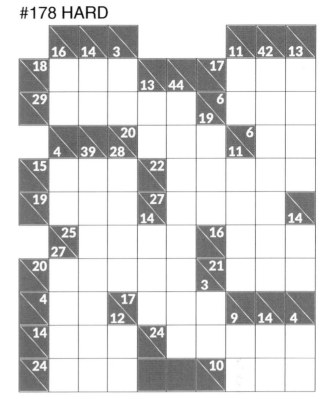

#179 MEDIUM

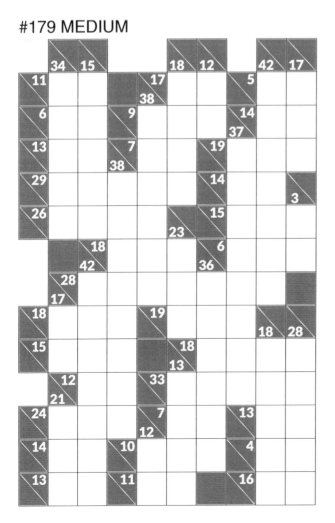

#180 HARD

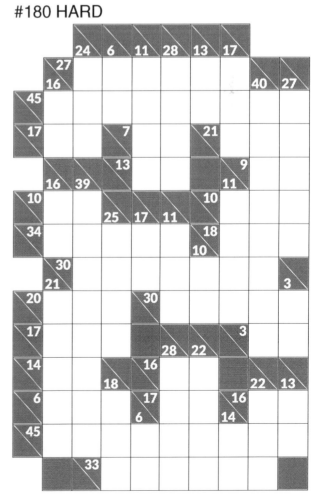

#181 MEDIUM

#182 HARD

#183 MEDIUM

#184 HARD

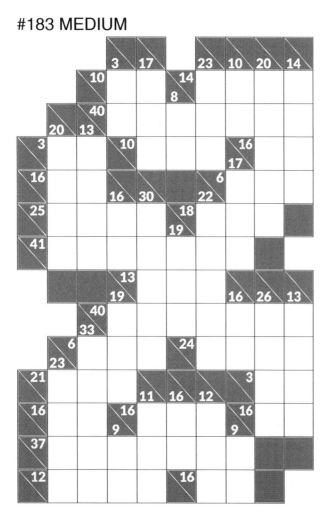

#185 MEDIUM

#186 HARD

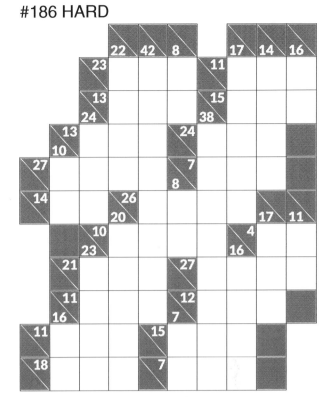

#187 MEDIUM

#188 HARD

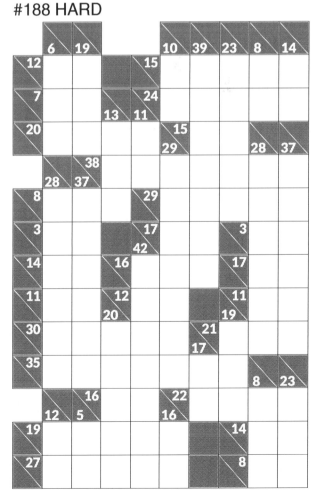

#189 MEDIUM

#190 HARD

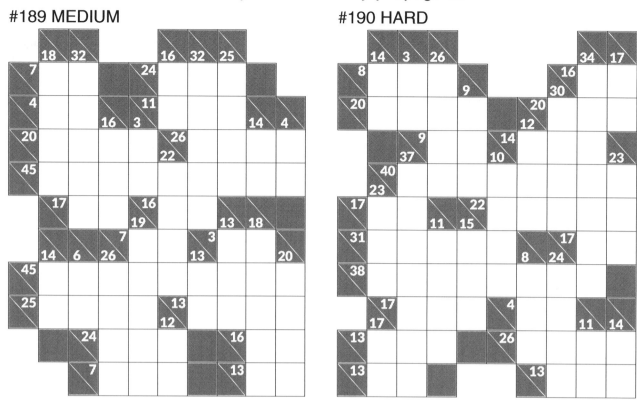

#191 MEDIUM

#192 HARD

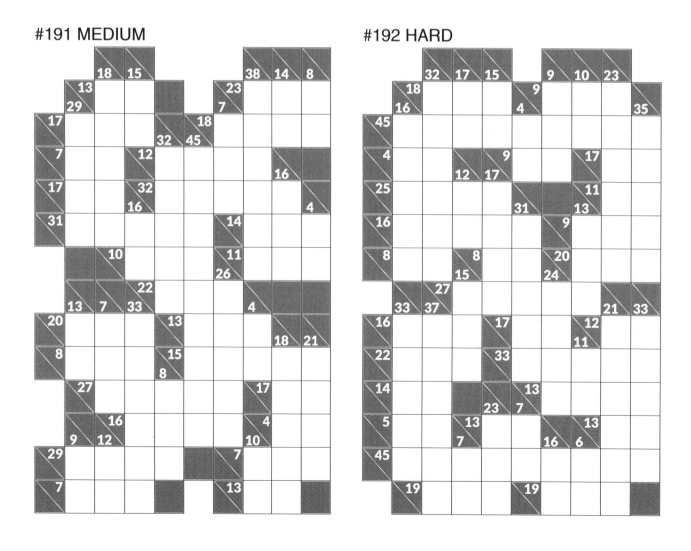

#193 MEDIUM

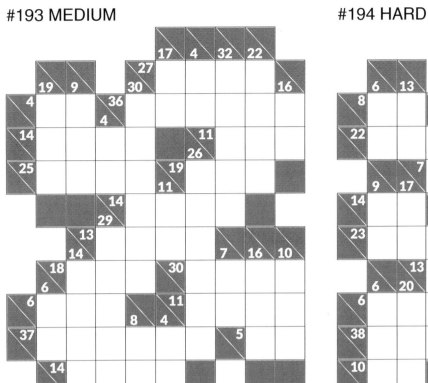

#194 HARD

#195 MEDIUM

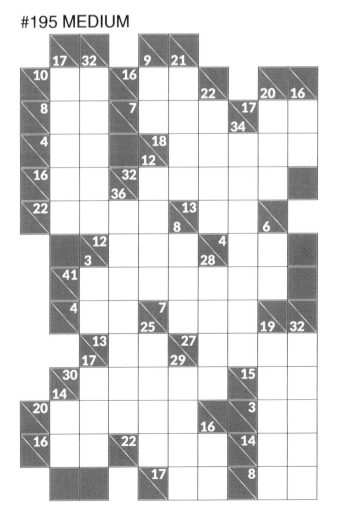

#196 HARD

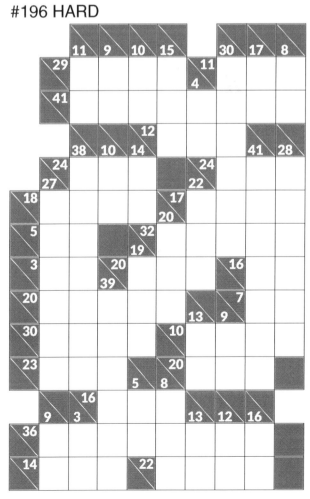

#197 MEDIUM

#198 HARD

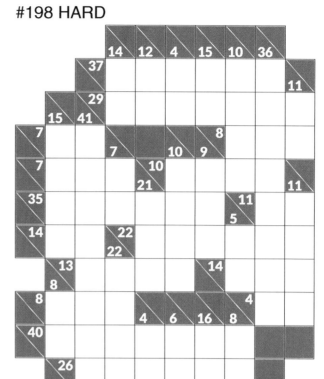

#199 MEDIUM

#200 HARD

#201 MEDIUM

#202 HARD

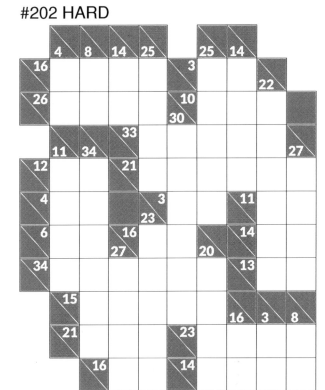

#203 MEDIUM

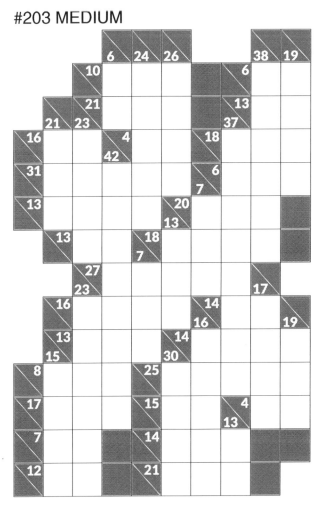

#204 HARD

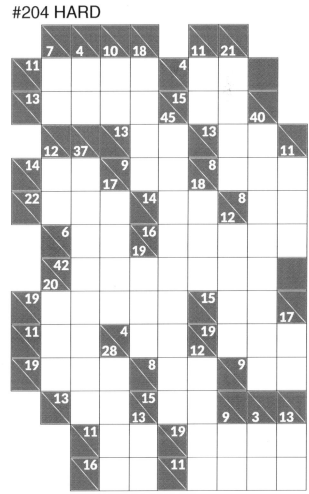

#205 MEDIUM

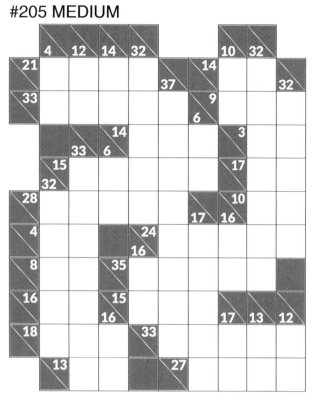

#206 HARD

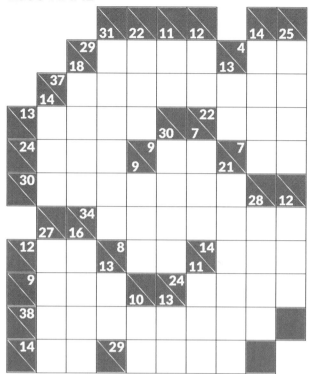

#207 MEDIUM

#208 HARD

#209 MEDIUM

#210 HARD

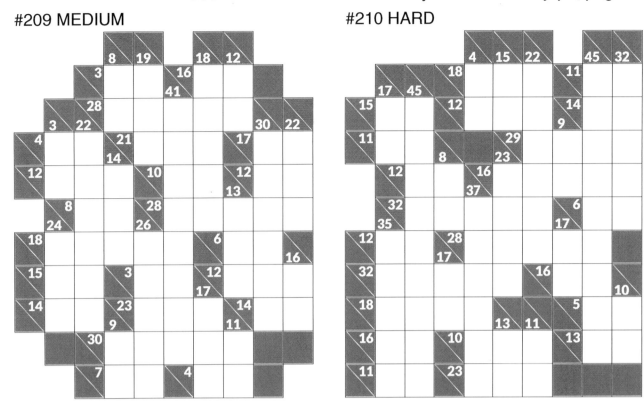

#211 MEDIUM

#212 HARD

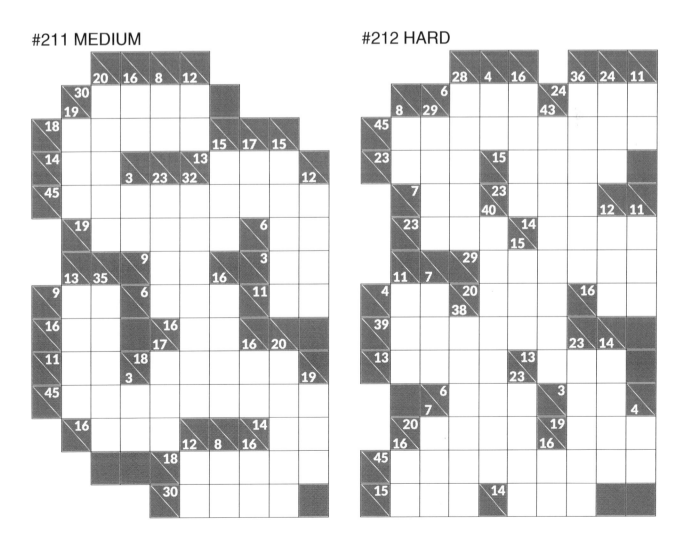

#213 MEDIUM

#214 HARD

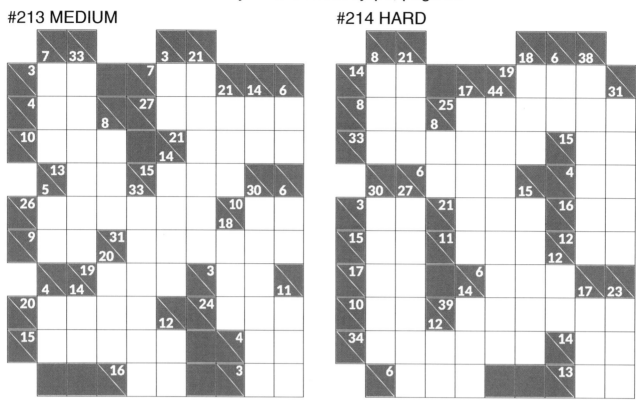

#215 MEDIUM

#216 HARD

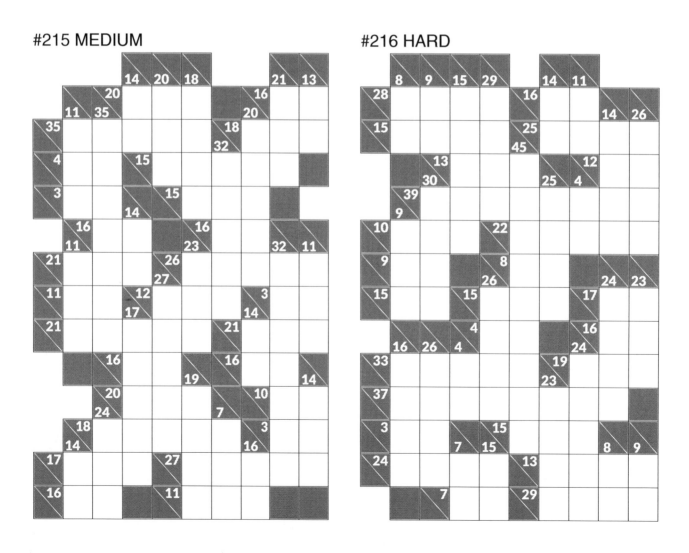

#217 MEDIUM

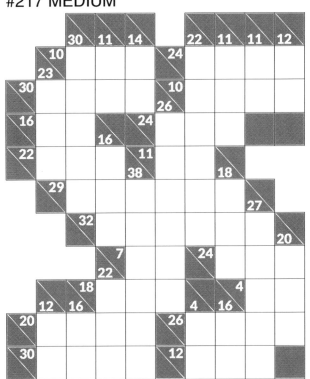

#218 HARD

#219 MEDIUM

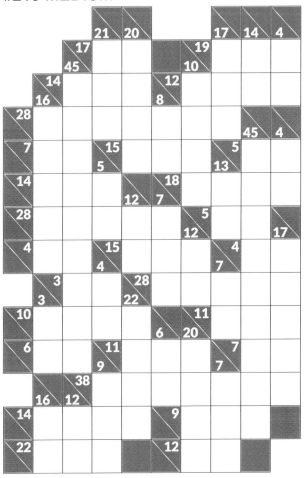

#220 HARD

#221 MEDIUM

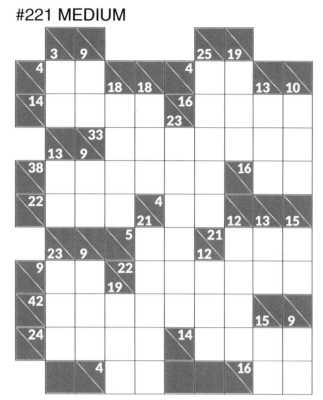

#222 HARD

#223 MEDIUM

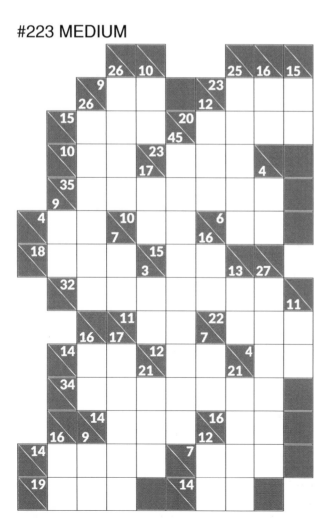

#224 HARD

#225 MEDIUM

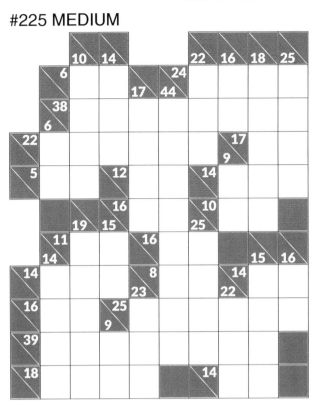

#226 HARD

#227 MEDIUM

#228 HARD

#229 MEDIUM

#230 HARD

#231 MEDIUM

#232 HARD

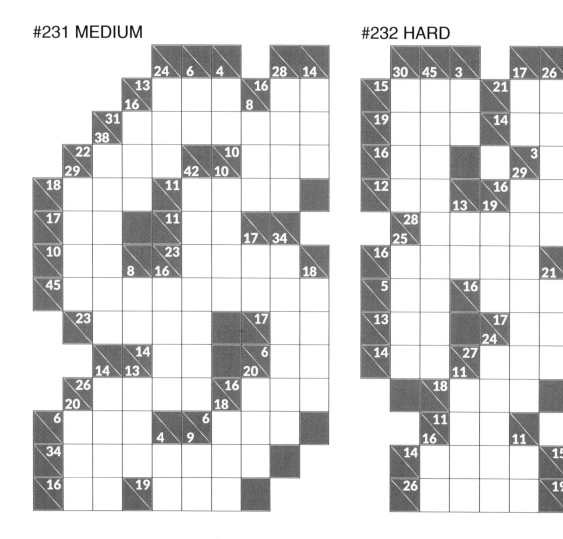

#233 MEDIUM

#234 HARD

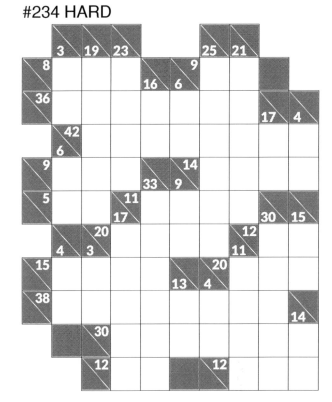

#235 MEDIUM

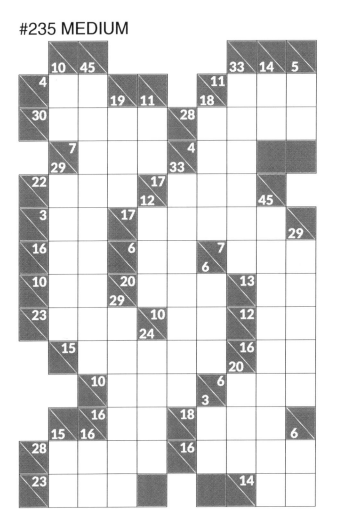

#236 HARD

#237 MEDIUM

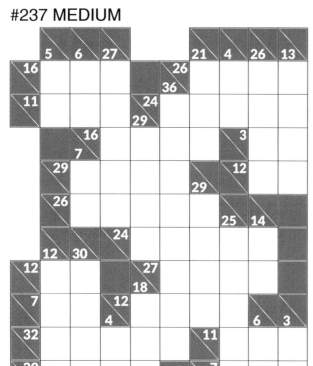

#238 HARD

#239 MEDIUM

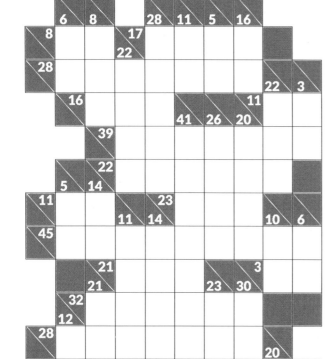

#240 HARD

#241 MEDIUM

#242 HARD

#243 MEDIUM

#244 HARD

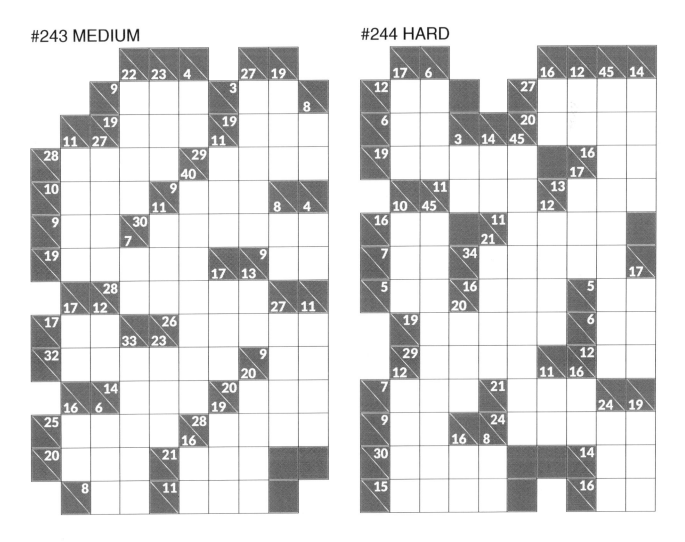

#245 MEDIUM

#246 HARD

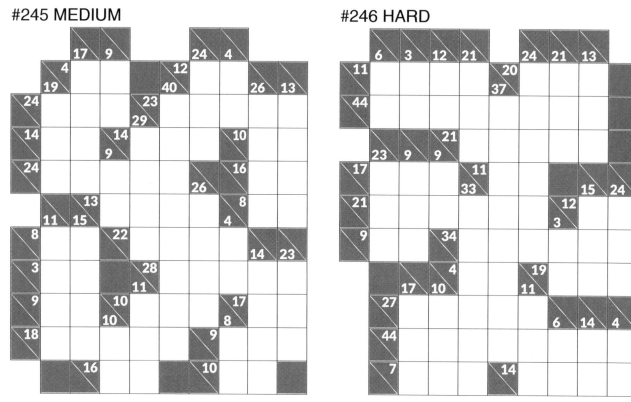

#247 MEDIUM

#248 HARD

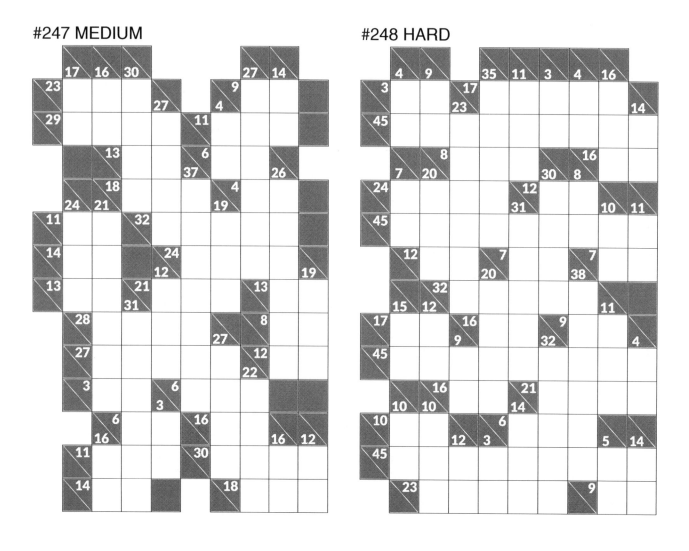

#249 MEDIUM

#250 HARD

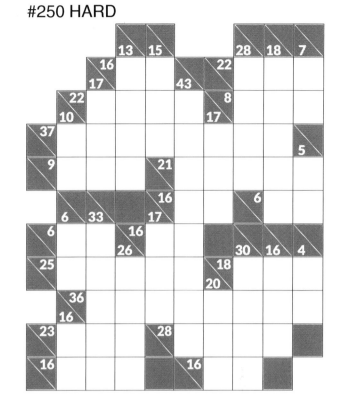

#251 MEDIUM

#252 HARD

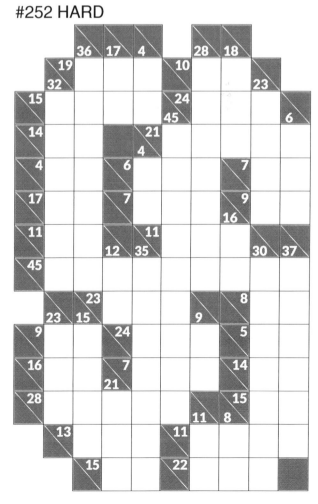

#253 MEDIUM

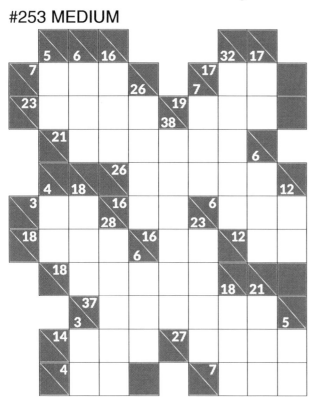

#254 HARD

#255 MEDIUM

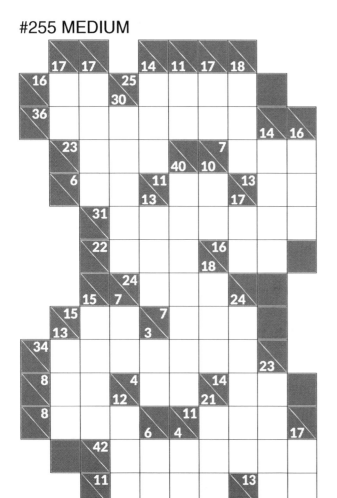

#256 HARD

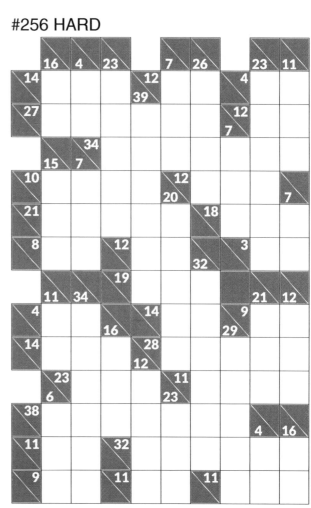

#257 MEDIUM

#258 HARD

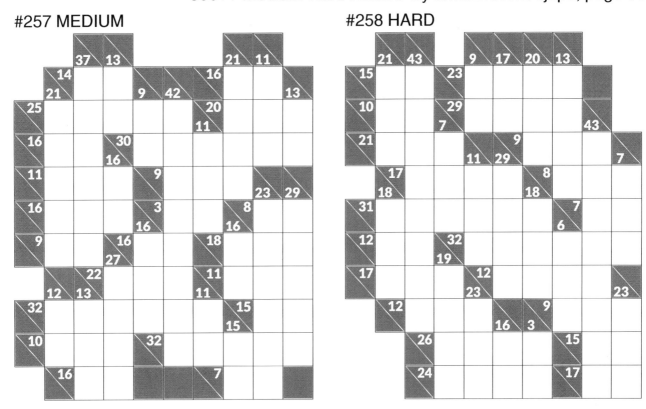

#259 MEDIUM

#260 HARD

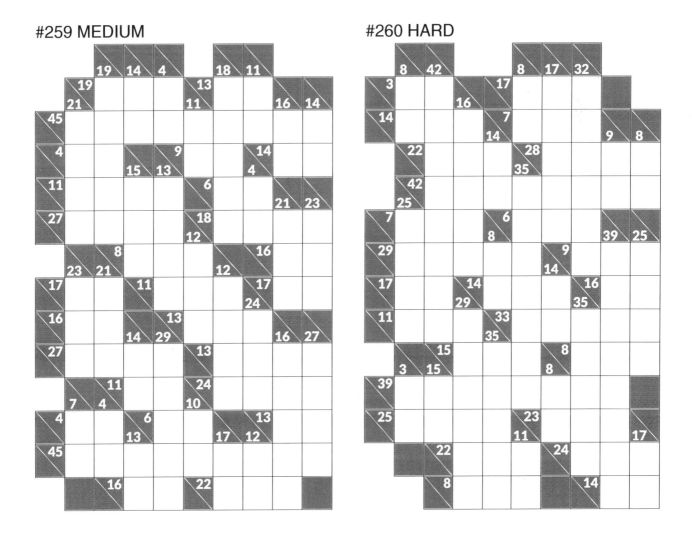

#261 MEDIUM

#262 HARD

#263 MEDIUM

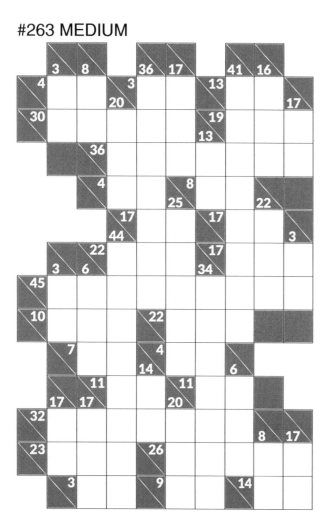

#264 HARD

#265 MEDIUM

#266 HARD

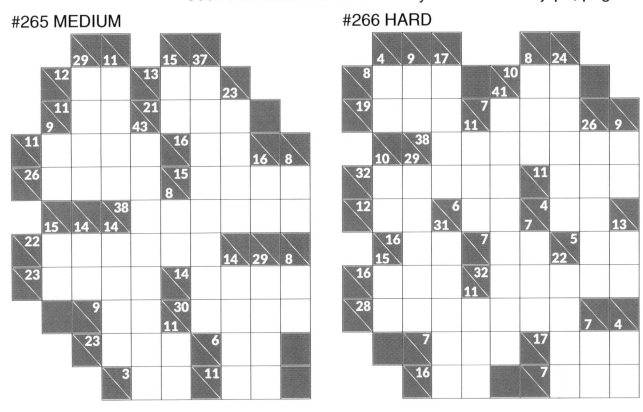

#267 MEDIUM

#268 HARD

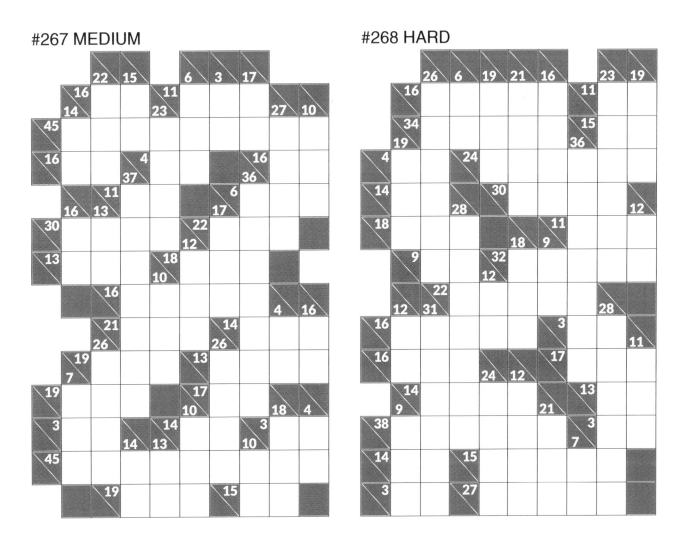

#269 MEDIUM

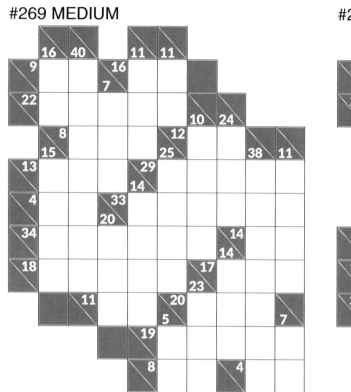

#270 HARD

#271 MEDIUM

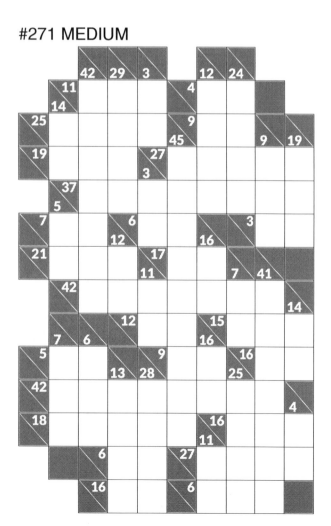

#272 HARD

#273 MEDIUM

#274 HARD

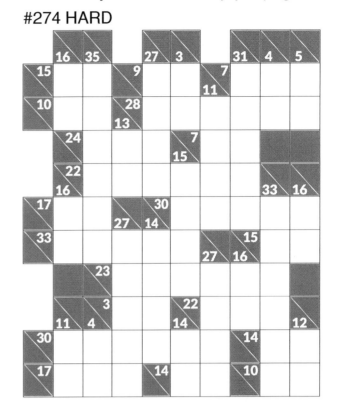

#275 MEDIUM

#276 HARD

#277 MEDIUM

#278 HARD

#279 MEDIUM

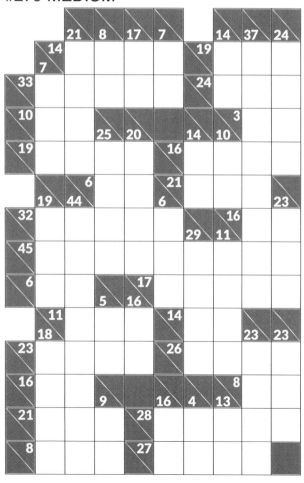

#280 HARD

#281 MEDIUM

#282 HARD

#283 MEDIUM

#284 HARD

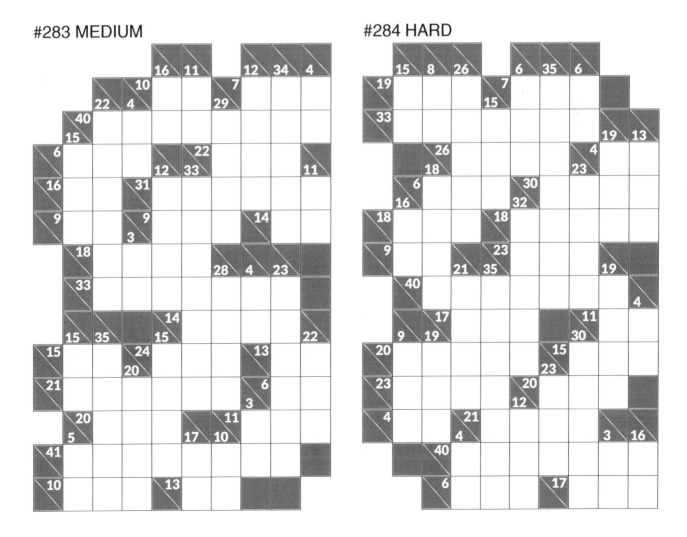

#285 MEDIUM

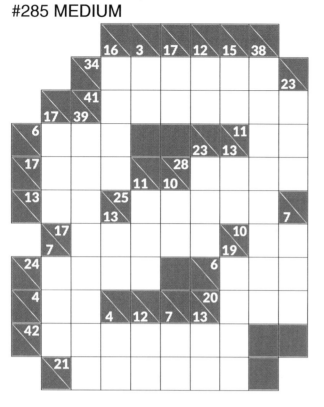

#286 HARD

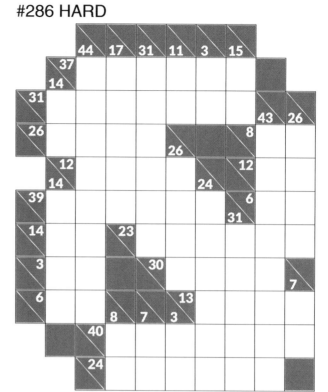

#287 MEDIUM

#288 HARD

#289 MEDIUM

#290 HARD

#291 MEDIUM

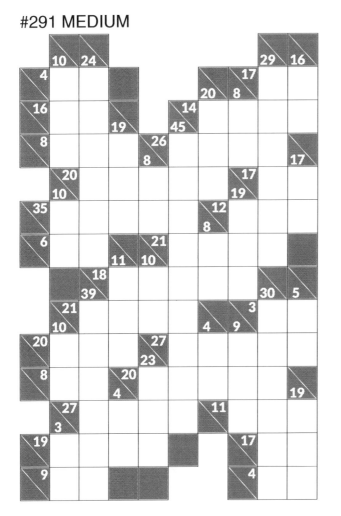

#292 HARD

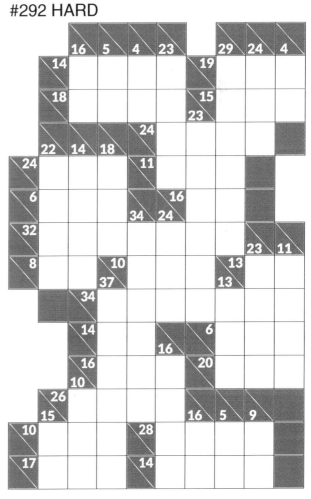

#293 MEDIUM

#294 HARD

#295 MEDIUM

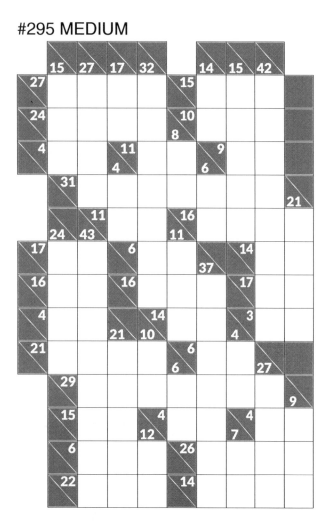

#296 HARD

#297 MEDIUM

#298 HARD

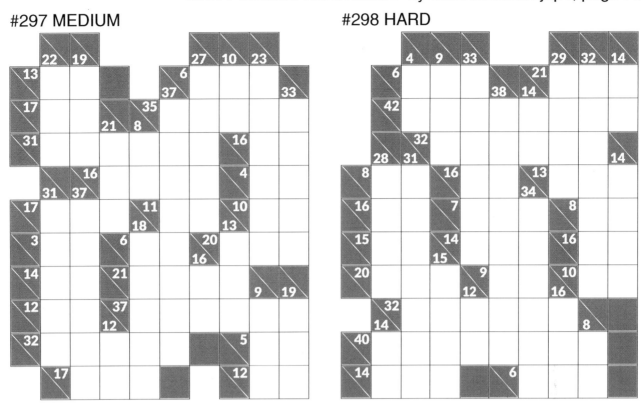

#299 MEDIUM

#300 HARD

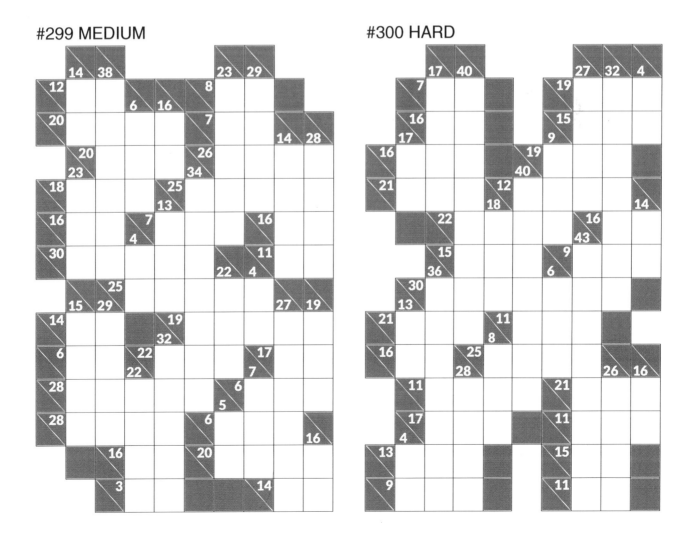

#301 MEDIUM

#302 HARD

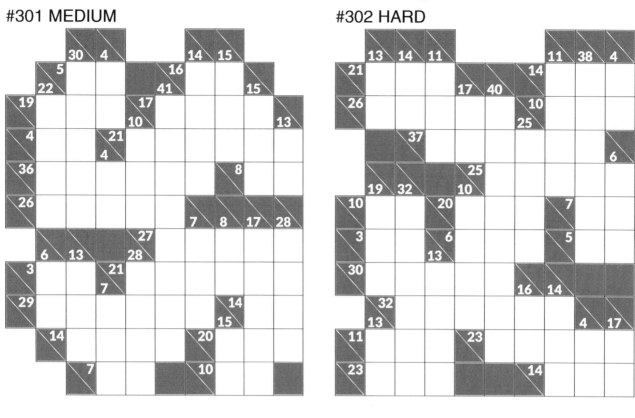

#303 MEDIUM

#304 HARD

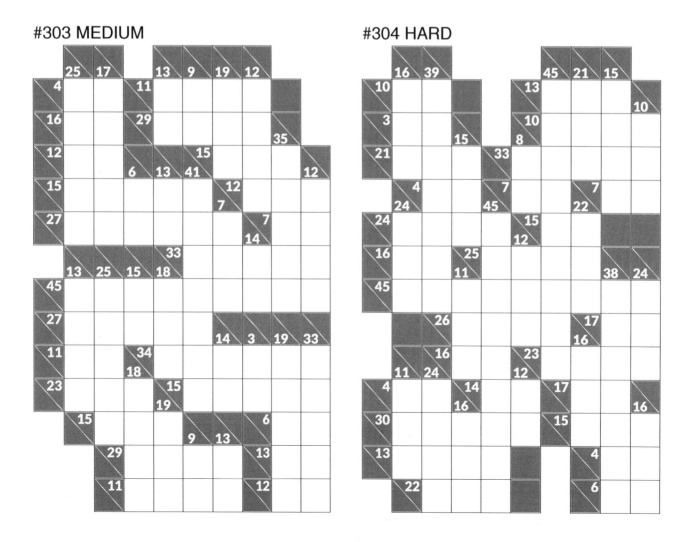

#305 MEDIUM

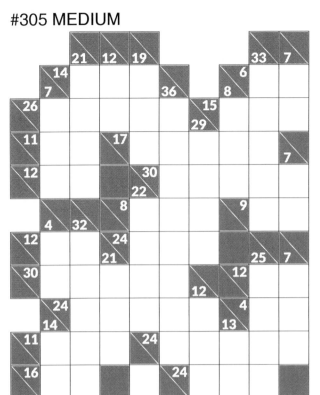

#306 HARD

#307 MEDIUM

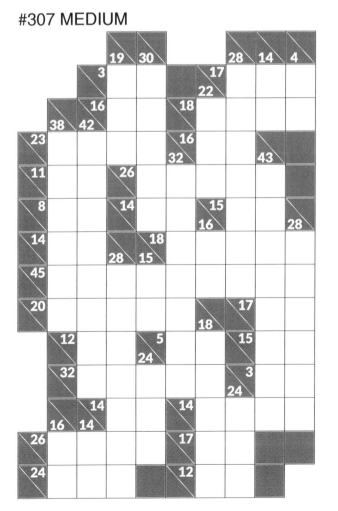

#308 HARD

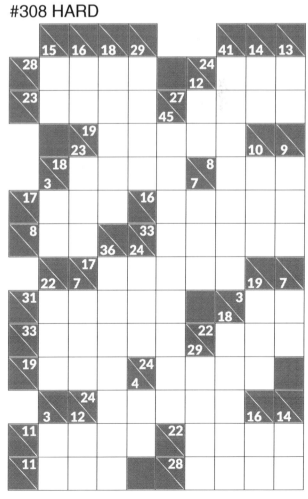

#309 MEDIUM

#310 HARD

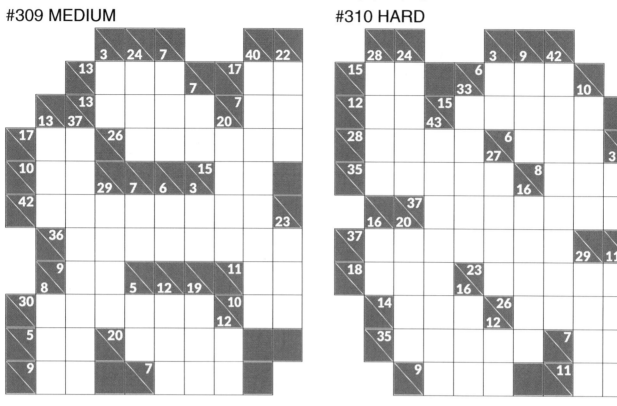

#311 MEDIUM

#312 HARD

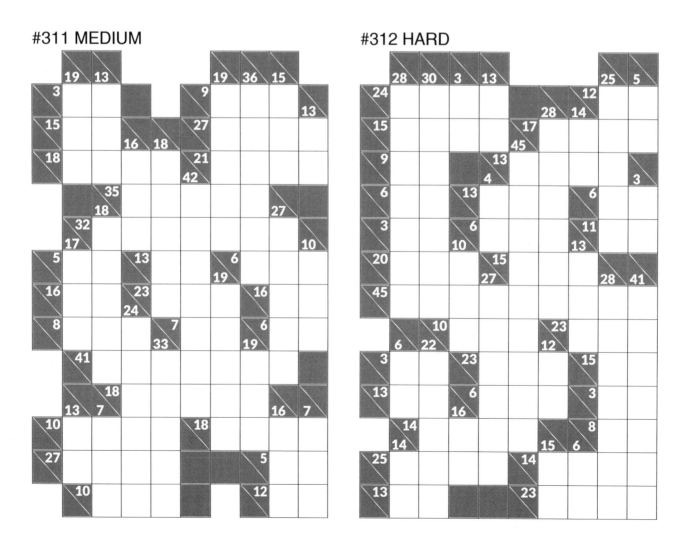

#313 MEDIUM

#314 HARD

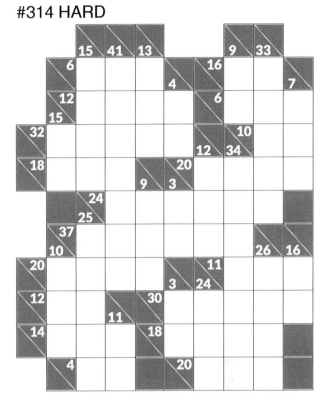

#315 MEDIUM

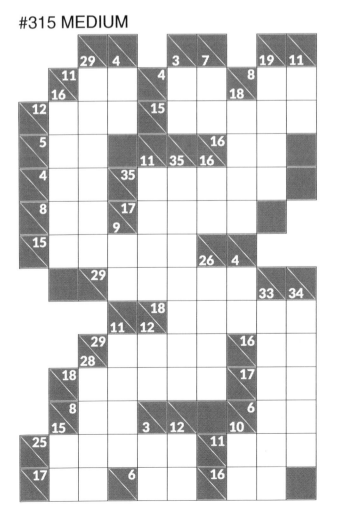

#316 HARD

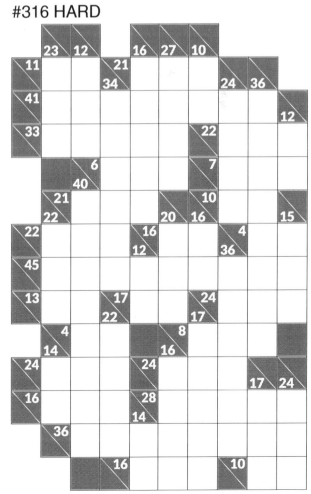

#317 MEDIUM

#318 HARD

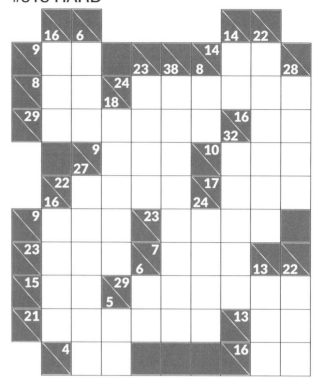

#319 MEDIUM

#320 HARD

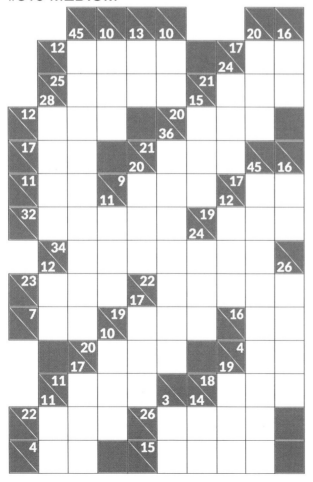

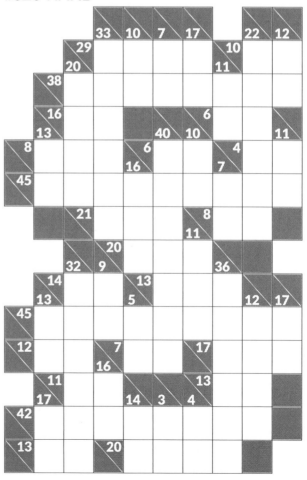

#321 MEDIUM

#322 HARD

#323 MEDIUM

#324 HARD

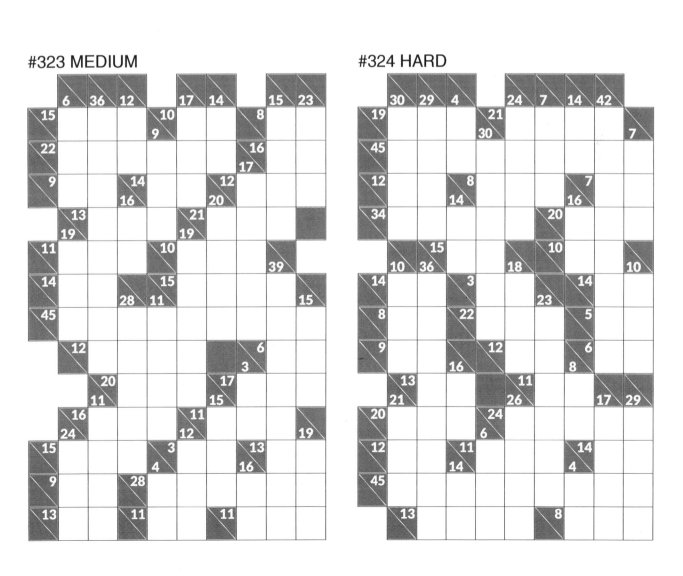

#325 MEDIUM

#326 HARD

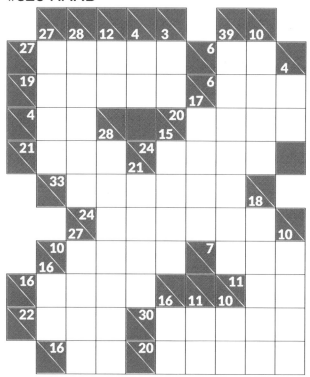

#327 MEDIUM

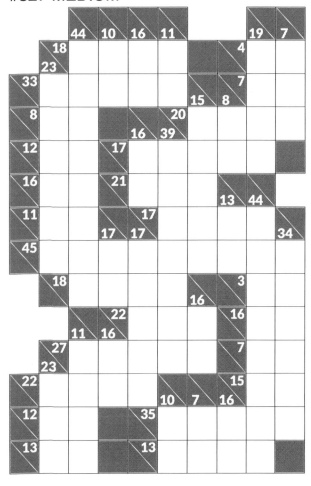

#328 HARD

#329 MEDIUM

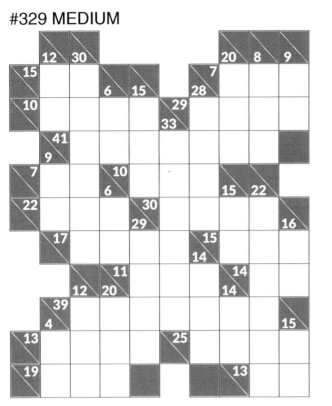

#330 HARD

#331 MEDIUM

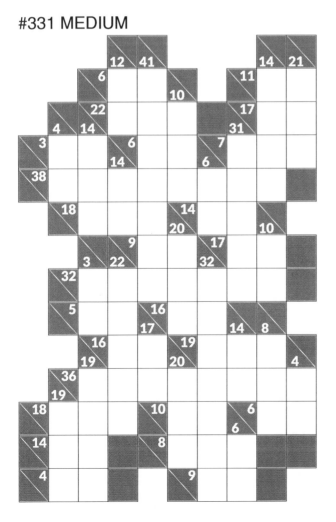

#332 HARD

#333 MEDIUM

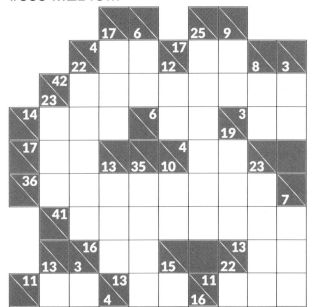

#334 HARD

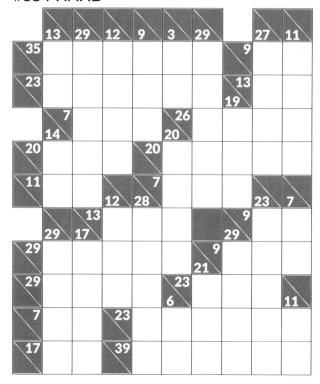

#335 MEDIUM

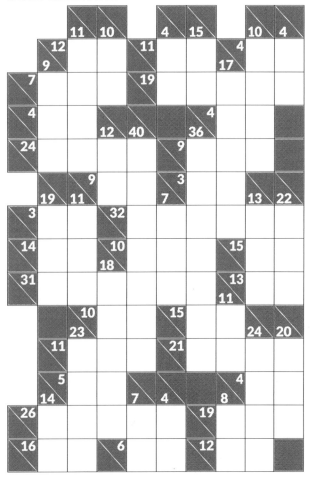

#336 HARD

#337 MEDIUM

#338 HARD

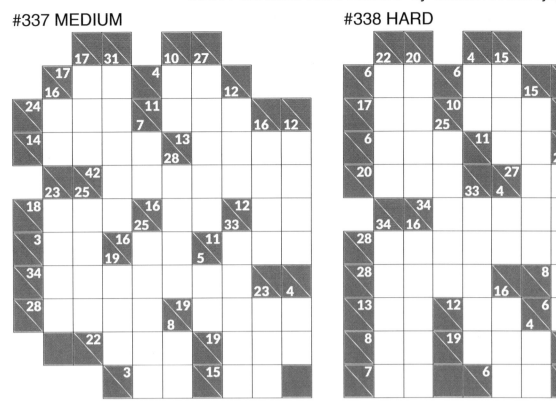

#339 MEDIUM

#340 HARD

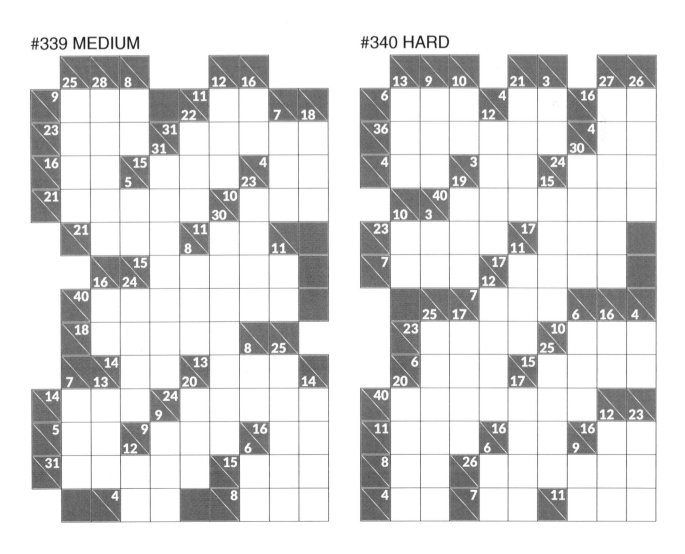

#341 MEDIUM

#342 HARD

#343 MEDIUM

#344 HARD

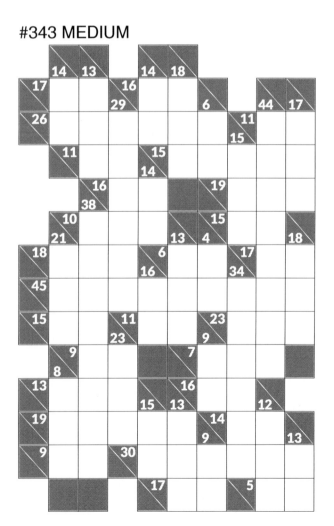

#345 MEDIUM

#346 HARD

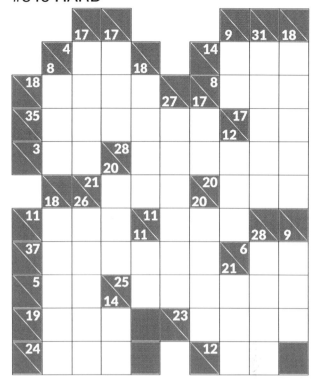

#347 MEDIUM

#348 HARD

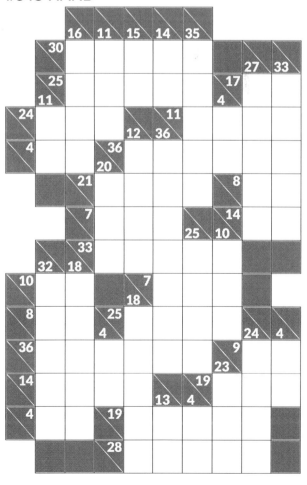

#349 MEDIUM

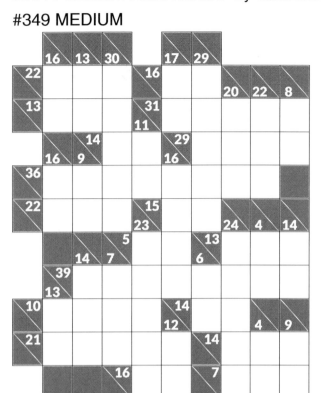

#350 HARD

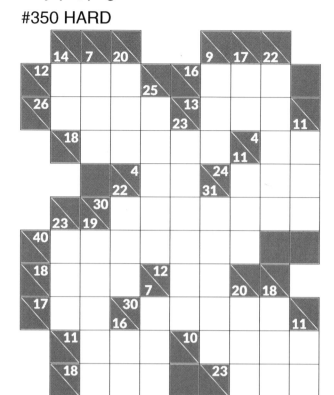

#351 MEDIUM

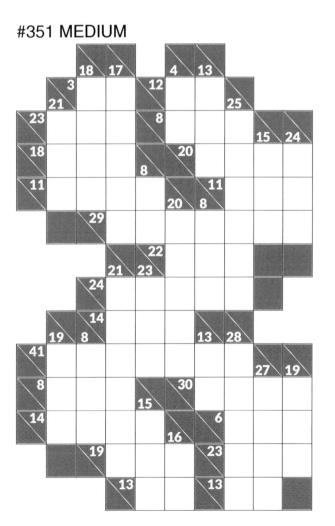

#352 HARD

#353 MEDIUM

#354 HARD

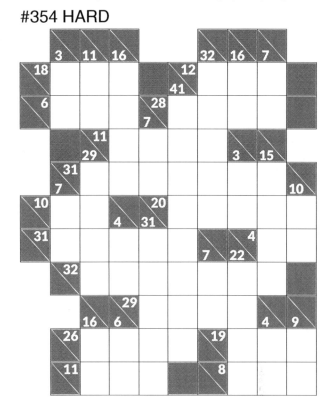

#355 MEDIUM

#356 HARD

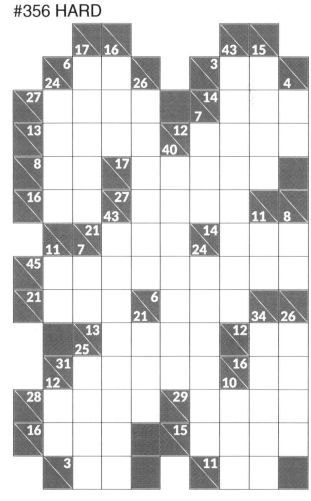

#357 MEDIUM

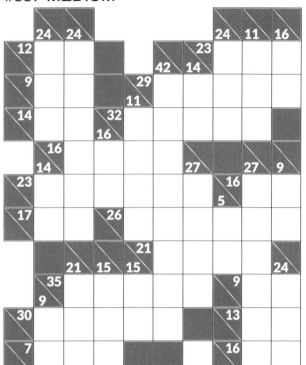

#358 HARD

#359 MEDIUM

#360 HARD

#361 MEDIUM

#362 HARD

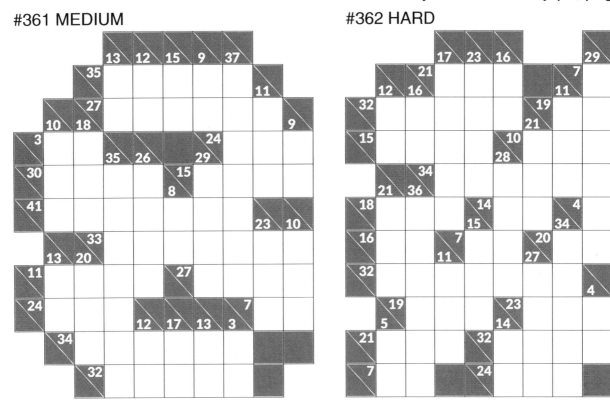

#363 MEDIUM

#364 HARD

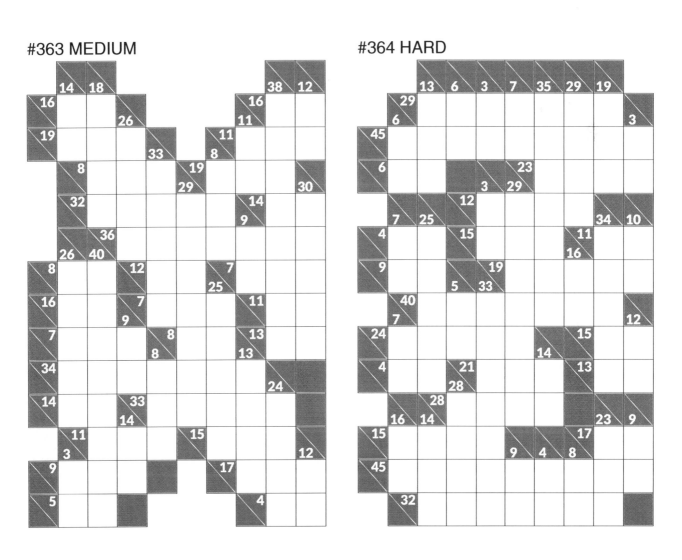

#365 MEDIUM

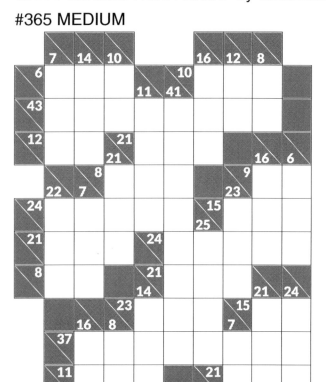

#366 HARD

#367 MEDIUM

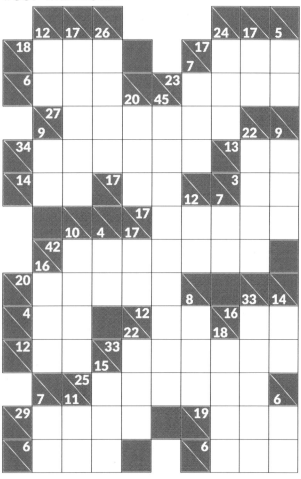

#368 HARD

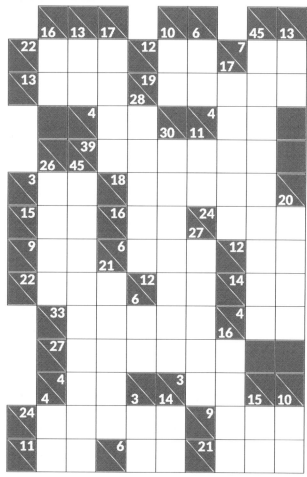

#369 MEDIUM

#370 HARD

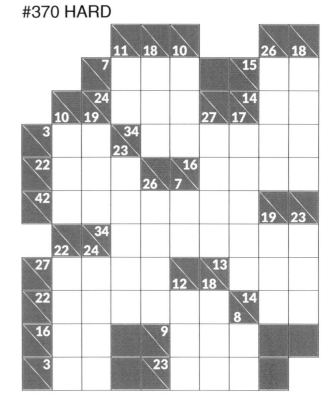

#371 MEDIUM

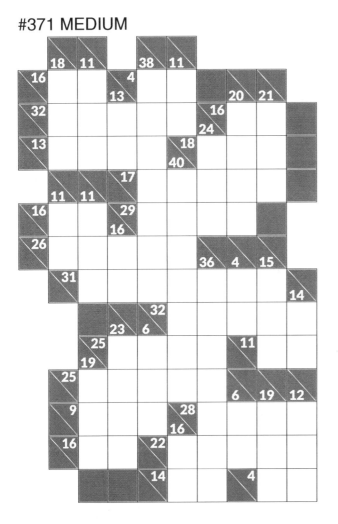

#372 HARD

#373 MEDIUM

#374 HARD

#375 MEDIUM

#376 HARD

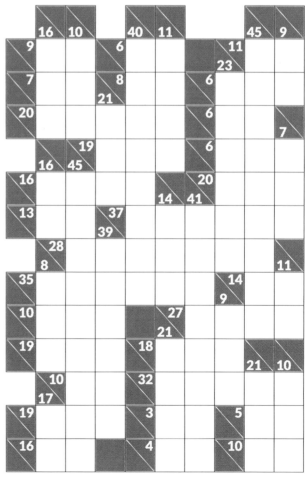

#377 MEDIUM

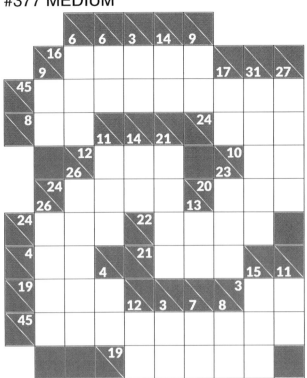

#378 HARD

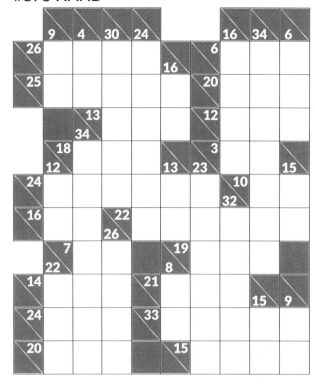

#379 MEDIUM

#380 HARD

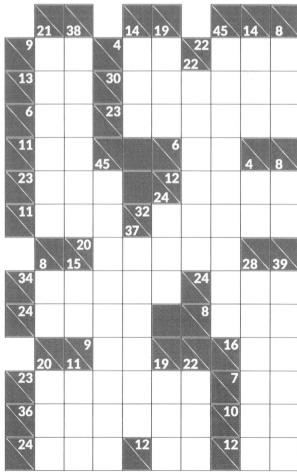

#381 MEDIUM

#382 HARD

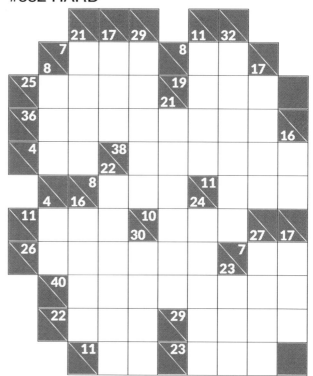

#383 MEDIUM

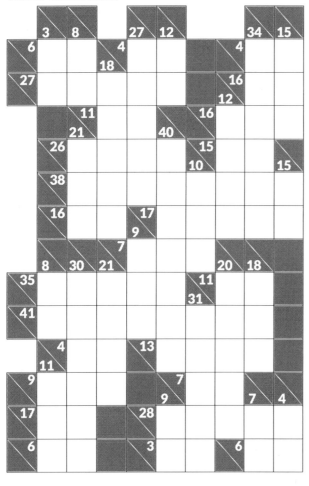

#384 HARD

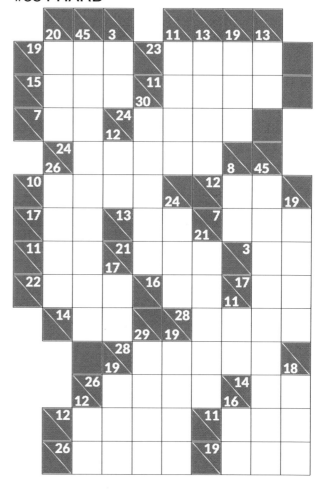

#385 MEDIUM

#386 HARD

#387 MEDIUM

#388 HARD

#389 MEDIUM

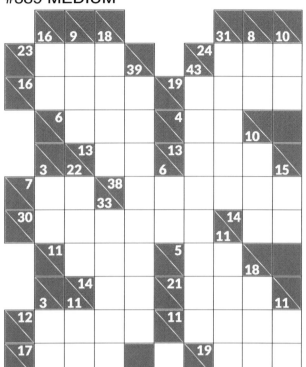

#390 HARD

#391 MEDIUM

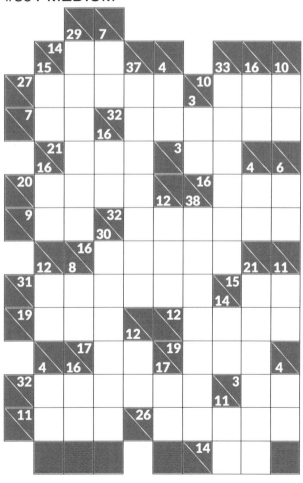

#392 HARD

#393 MEDIUM

#394 HARD

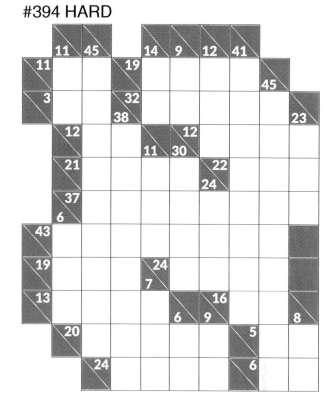

#395 MEDIUM

#396 HARD

#397 MEDIUM

#398 HARD

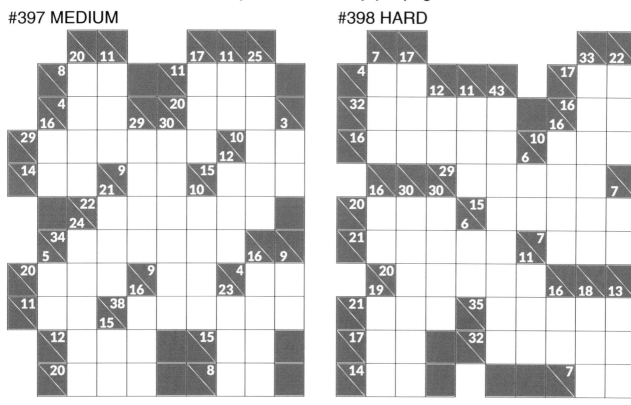

#399 MEDIUM

#400 HARD

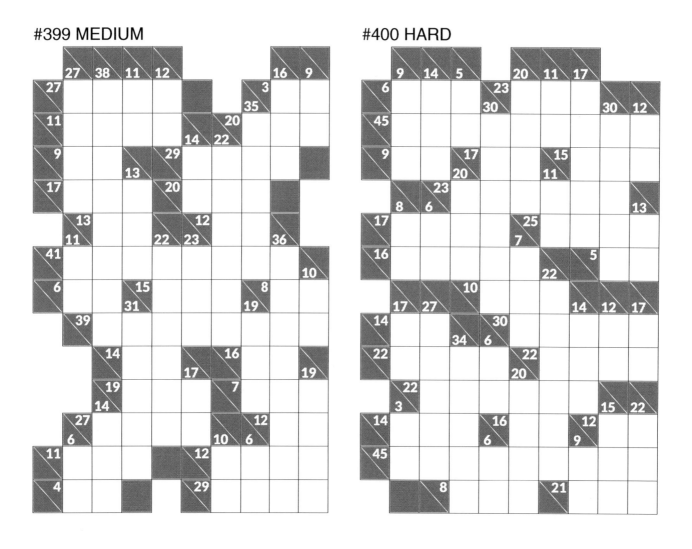

#401 MEDIUM

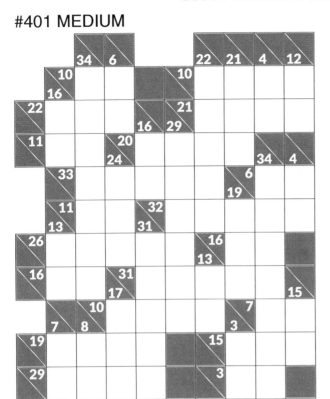

#402 HARD

#403 MEDIUM

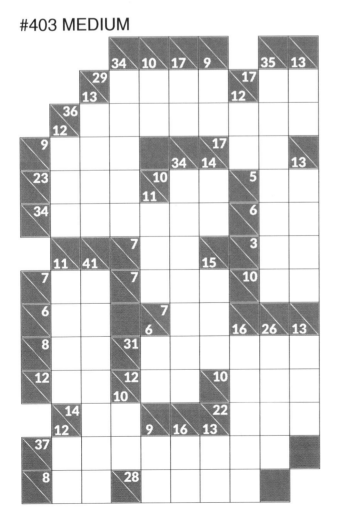

#404 HARD

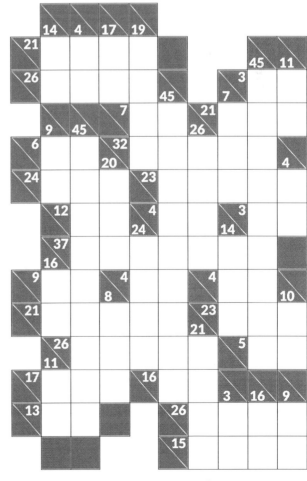

#405 MEDIUM

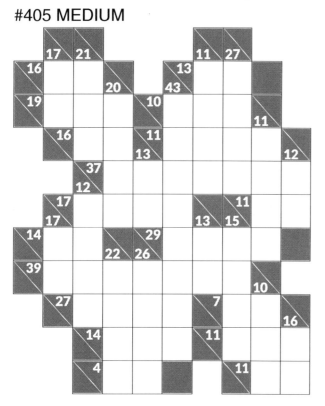

#406 HARD

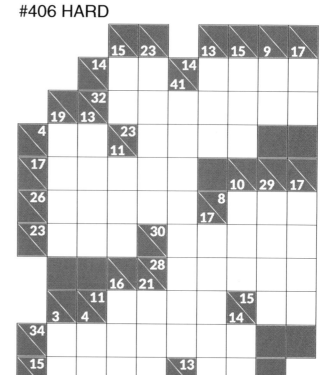

#407 MEDIUM

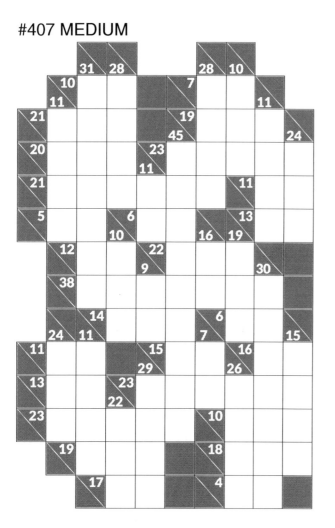

#408 HARD

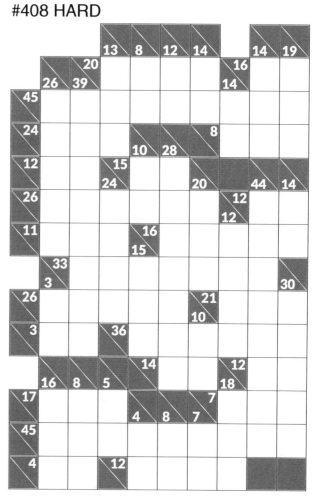

#409 MEDIUM

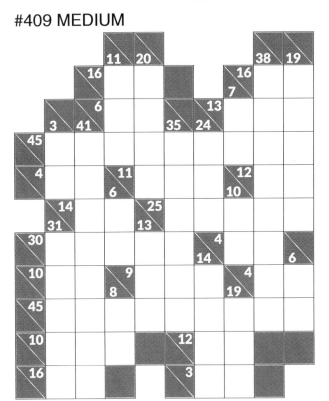

#410 HARD

#411 MEDIUM

#412 HARD

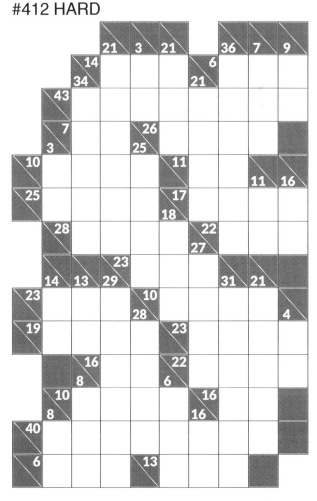

#413 MEDIUM

#414 HARD

#415 MEDIUM

#416 HARD

#417 MEDIUM

#418 HARD

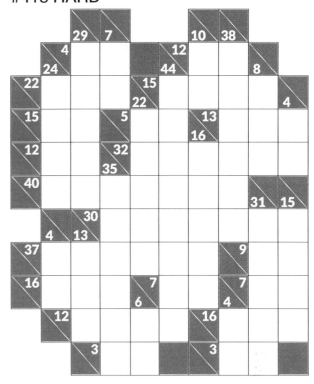

#419 MEDIUM

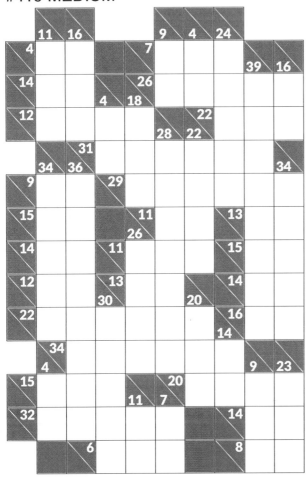

#420 HARD

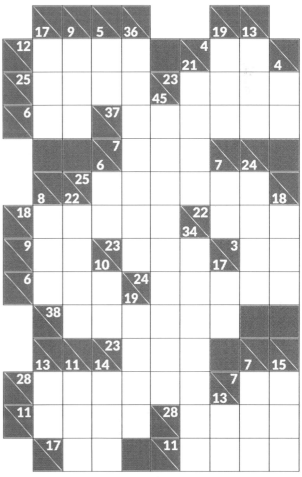

#421 MEDIUM

#422 HARD

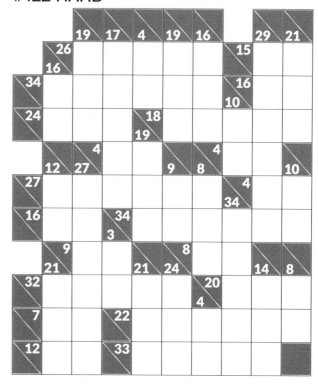

#423 MEDIUM

#424 HARD

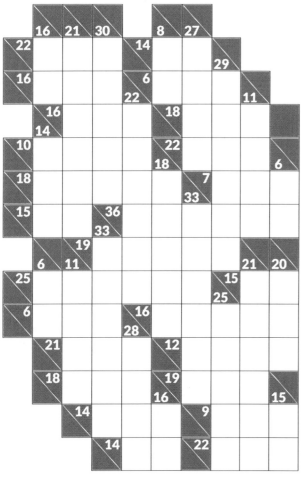

#425 MEDIUM

#426 HARD

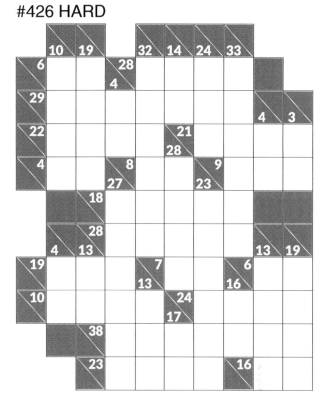

#427 MEDIUM

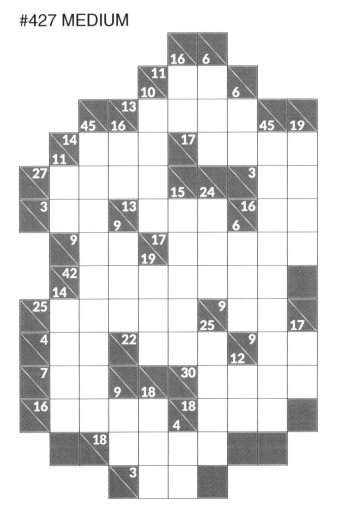

#428 HARD

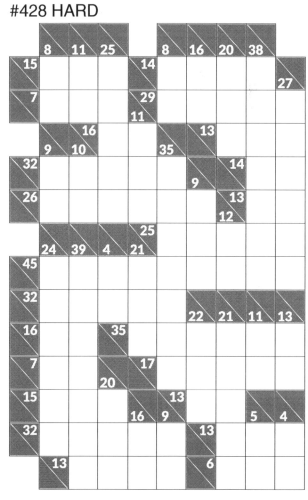

#429 MEDIUM

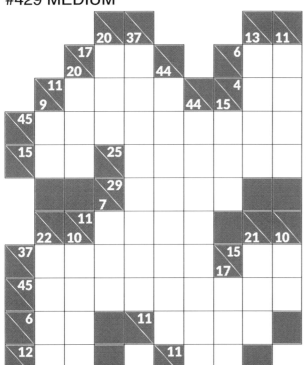

#430 HARD

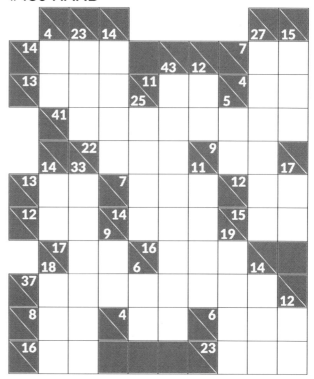

#431 MEDIUM

#432 HARD

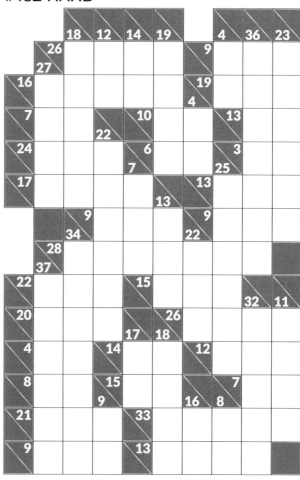

#433 MEDIUM

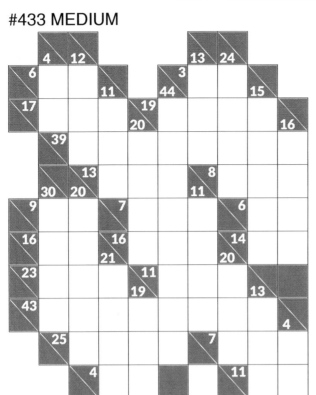

#434 HARD

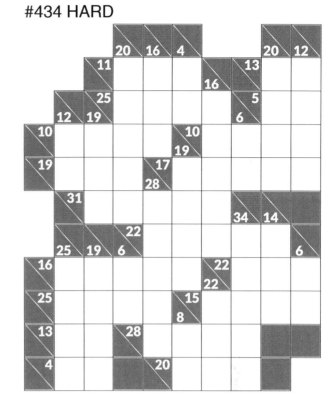

#435 MEDIUM

#436 HARD

#437 MEDIUM

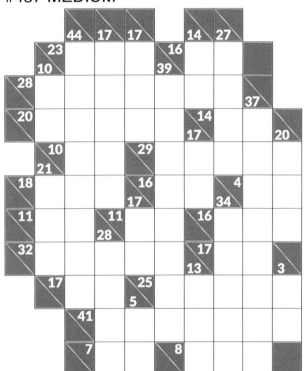

#438 HARD

#439 MEDIUM

#440 HARD

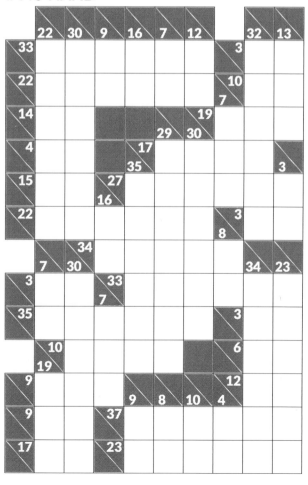

#441 MEDIUM

#442 HARD

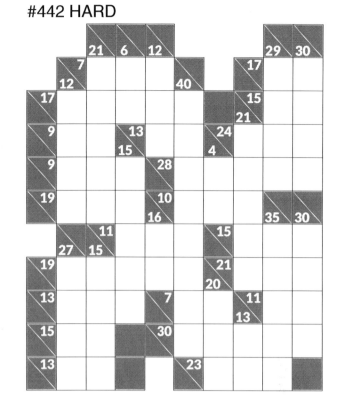

#443 MEDIUM

#444 HARD

#445 MEDIUM

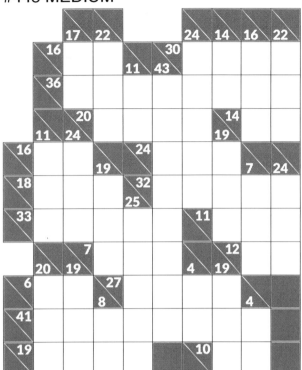

#446 HARD

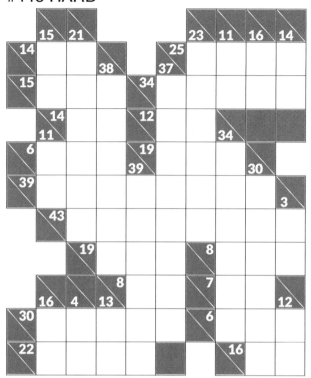

#447 MEDIUM

#448 HARD

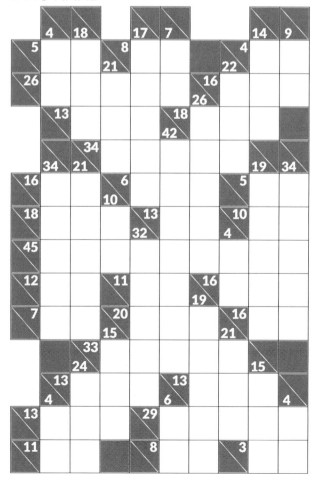

#449 MEDIUM

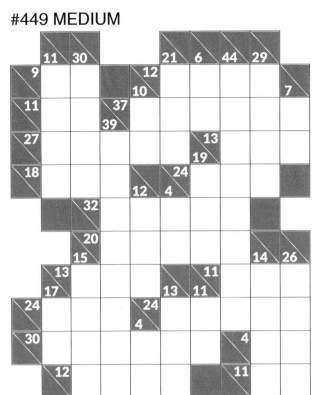

#450 HARD

#451 MEDIUM

#452 HARD

#453 MEDIUM

#454 HARD

#455 MEDIUM

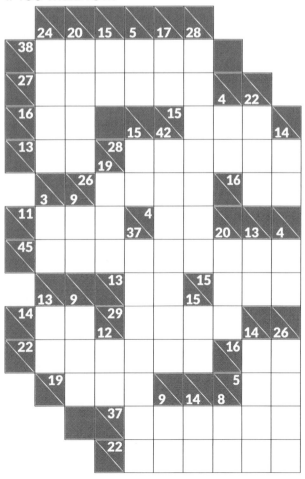

#456 HARD

#457 MEDIUM

#458 HARD

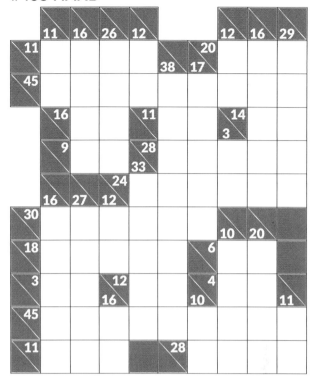

#459 MEDIUM

#460 HARD

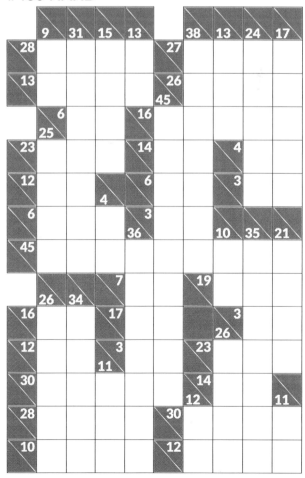

#461 MEDIUM

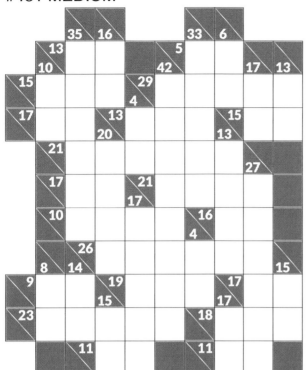

#462 HARD

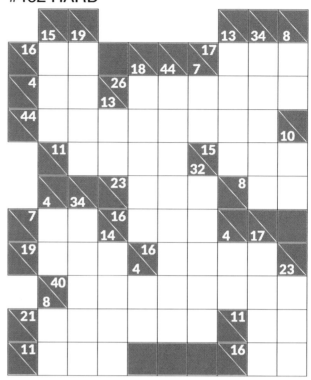

#463 MEDIUM

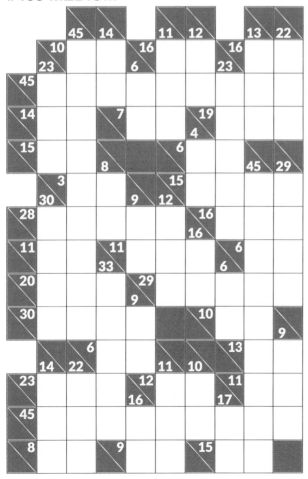

#464 HARD

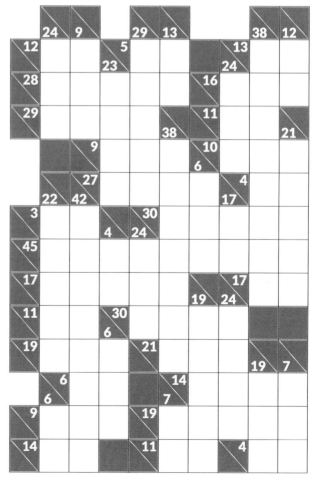

#465 MEDIUM

#466 HARD

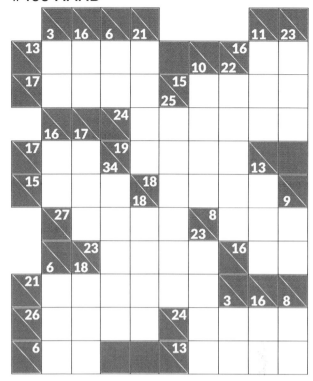

#467 MEDIUM

#468 HARD

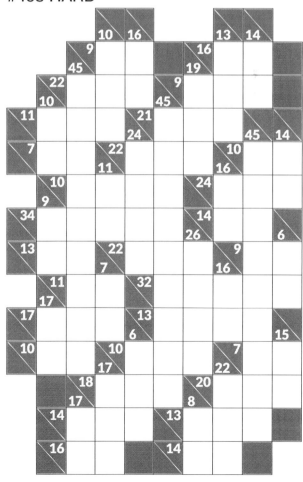

#469 MEDIUM

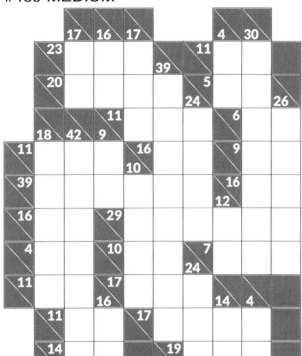

#470 HARD

#471 MEDIUM

#472 HARD

#473 MEDIUM

#474 HARD

#475 MEDIUM

#476 HARD

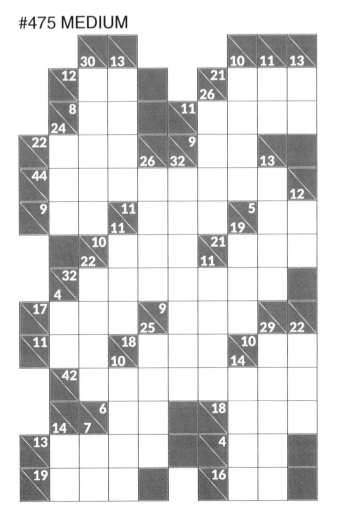

#477 MEDIUM

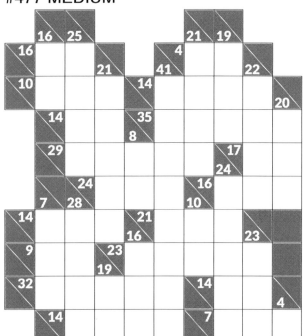

#478 HARD

#479 MEDIUM

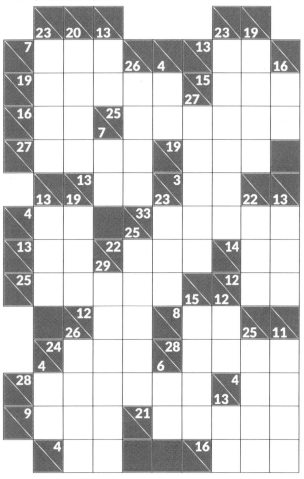

#480 HARD

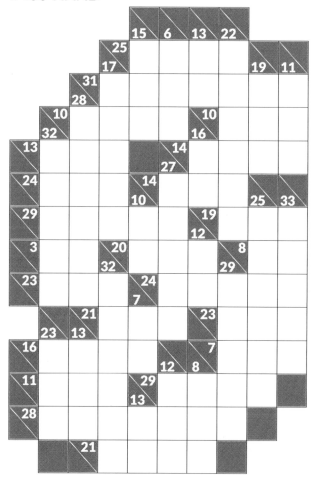

#481 MEDIUM

#482 HARD

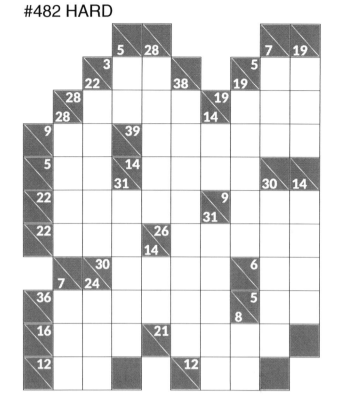

#483 MEDIUM

#484 HARD

#485 MEDIUM

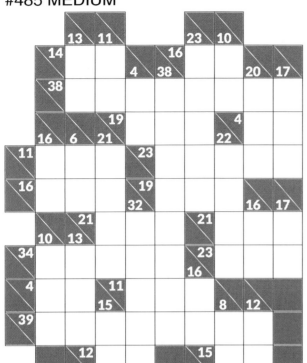

#486 HARD

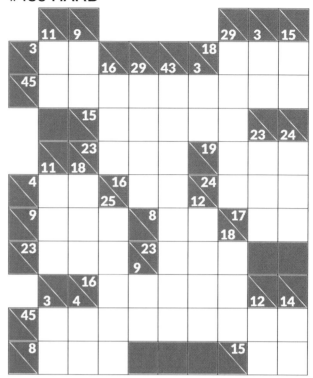

#487 MEDIUM

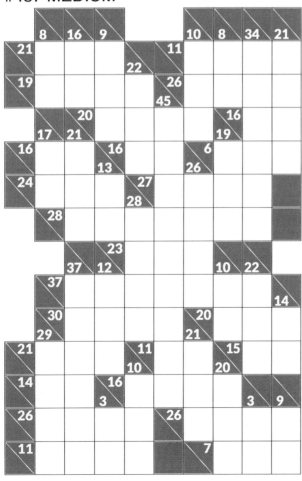

#488 HARD

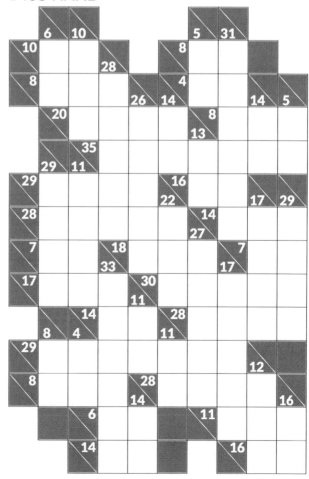

#489 MEDIUM

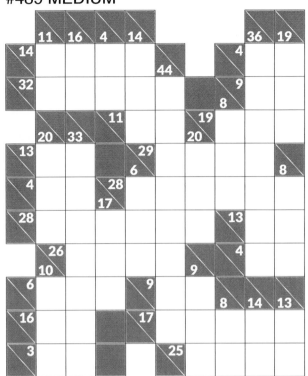

#490 HARD

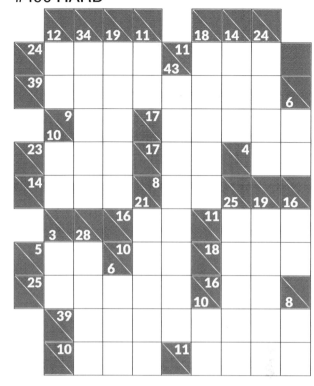

#491 MEDIUM

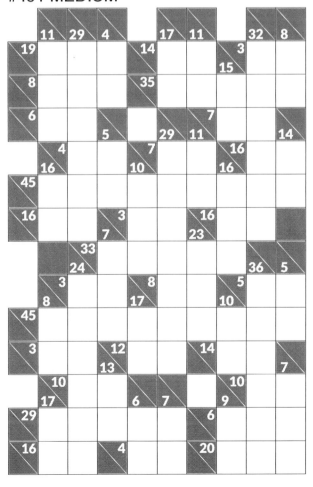

#492 HARD

#493 MEDIUM

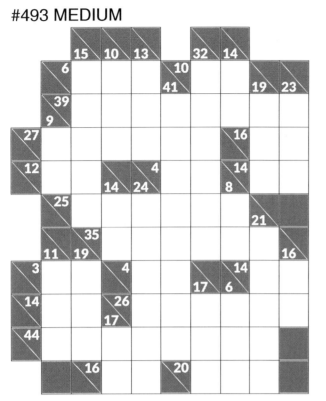

#494 HARD

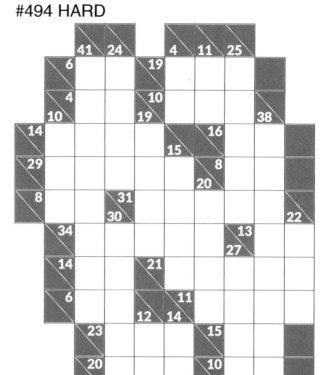

#495 MEDIUM

#496 HARD

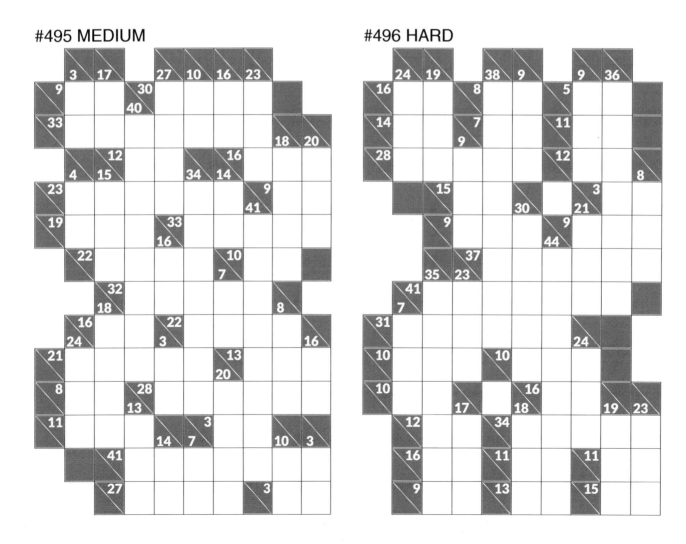

#497 MEDIUM

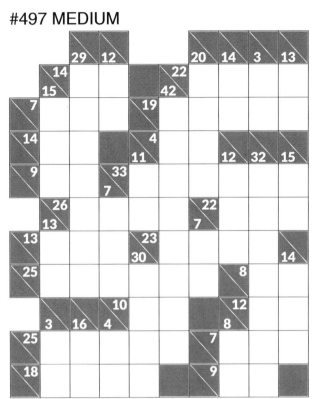

#498 HARD

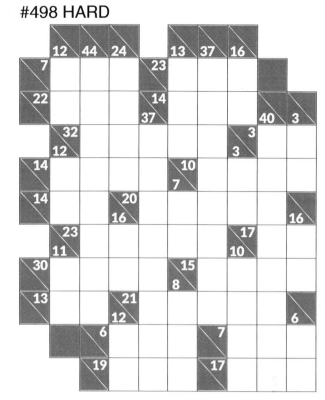

#499 MEDIUM

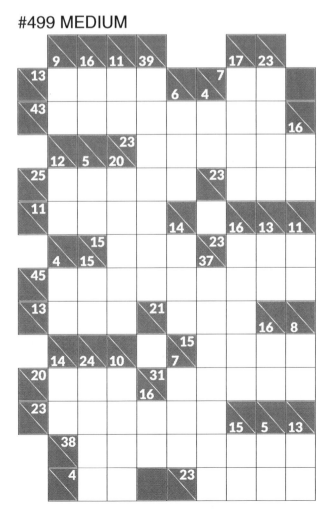

#500 HARD

#501 MEDIUM

#502 HARD

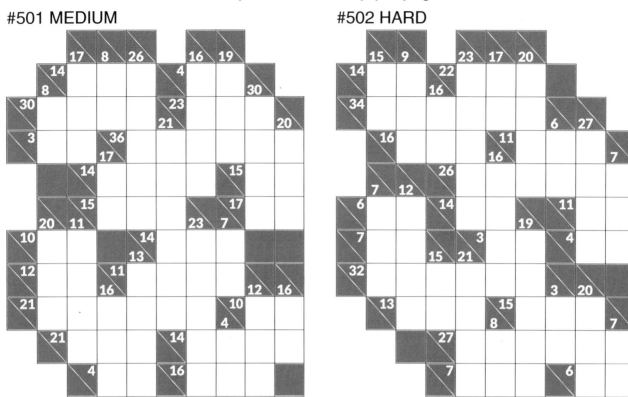

#503 MEDIUM

#504 HARD

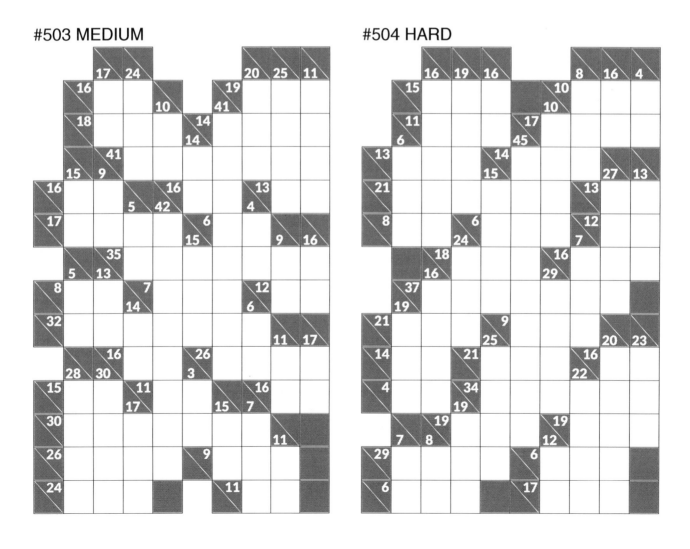

#505 MEDIUM

#506 HARD

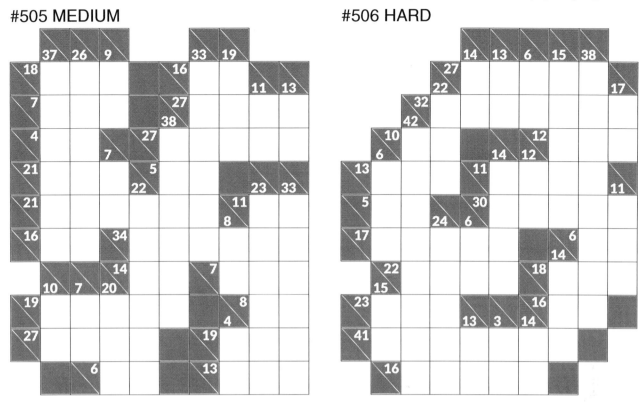

#507 MEDIUM

#508 HARD

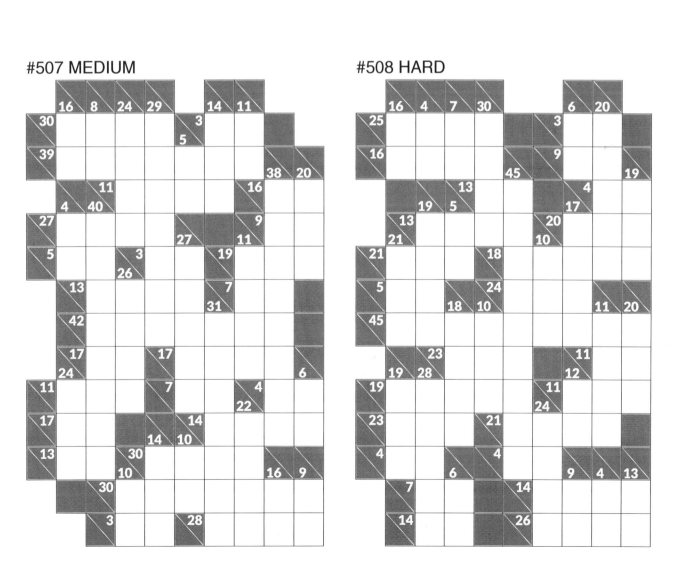

#509 MEDIUM

#510 HARD

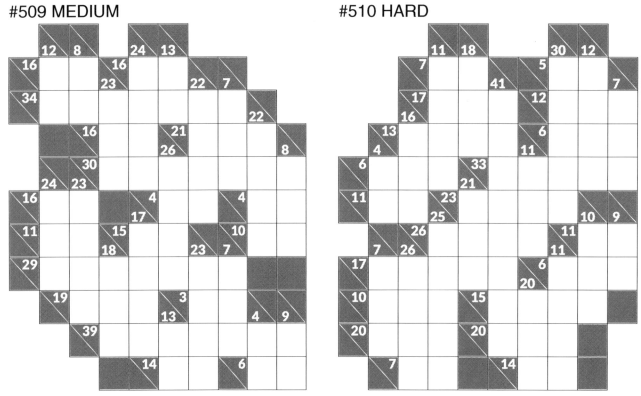

#511 MEDIUM

#512 HARD

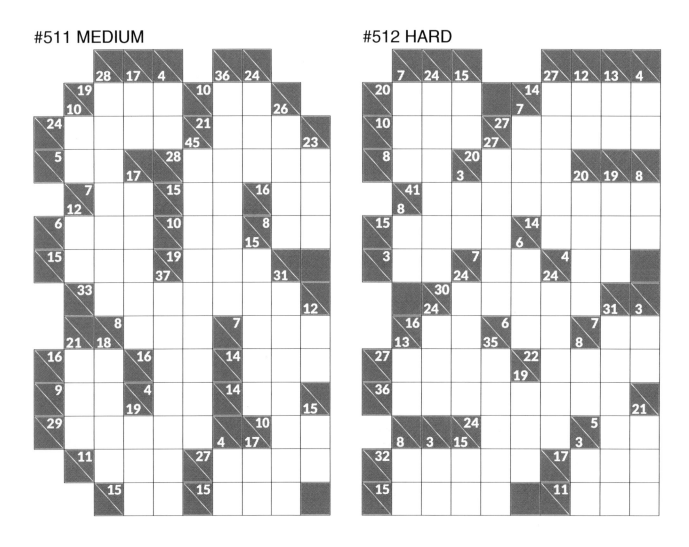

#1

		3	1		6	9		
	7	9	2		8	1	9	
2	7	1	4		1	2	4	3
1	9	3	8		6	9	8	
4	8		6	9	8	7	3	
	5	9	1	7	3		2	1
	4	7	5		9	8	7	5
7	1	3	2		4	2	5	3
9	6	8		9	7	4		
	2	5		6	5			

#2

	3	1	7	5	2			3	1
	7	8	9	6	1		2	7	5
			7	4	2	1	5	6	
3	7				6	8	9		
1	2		3	7	5	4	8	9	
6	5	1	8	9	7		2	7	
	1	4	2				1	3	
9	8	7	6	5	1				
1	4	2		7	2	6	1	4	
7	3			9	4	8	3	6	

#3

8	9	5		2	5	1	3	
5	2	1	7	3	9	4	6	8
	2	9	8			8	9	
		3	1	6	2			
1	2			3	1	7		
2	4	1		6	5	4	9	
1	3		8	9	7		4	9
	9	7	3	1		1	2	7
	8	9	6			9	8	
	3	1	4	5				
1	5			7	9	3		
5	3	4	8	2	6	1	7	9
4	2	3	1		8	9	6	

#4

	1	4		2	4		1	5	2
	5	6		4	3	1	2	7	9
	7	9			5	7	9		
1	3	8		5	4	9			
8	2		2	1	3		7	9	
5	8	4	9	2		7	9	4	
	5	8	4	7	2	3	1		
8	9	7		9	7	5	6	8	
3	1		9	8	5		4	7	
		3	8	6		4	2	1	
		3	8	7			8	5	
9	1	7	5	6	8		3	2	
6	7	9		4	9		8	9	

#5

		6	7			1	5	2
4	6	2	1			5	8	3
2	8	1		1	8	7	9	
1	9	8	2	7	6	4	3	5
			7	9			7	9
9	3			2	8			
5	1	7	6	3	9	8	4	2
	2	3	8	4		4	3	1
9	6	8			3	7	9	4
7	5	9			1	2		

#6

5	4			9	1	7	5	2
9	8		2	8	3	9	6	1
		1	2	3	7			
1	3	6	8		1	2		
2	9		7	8	9	5		
			5	1	2	3		6	1
		8	9		6	7	8	3
			3	2	1	5		
5	6	3	1	9	4		7	5
9	5	7	3	8			9	8

#7

1	3		7	2		2	7	9
9	7		9	8	2	1	4	7
	1	2	4		5	7		
	2	9			8	9	7	
		3	4		1	8	4	
	3	1	7	9			2	9
9	1	4	6	5	2	8	3	7
3	4			2	3	4	1	
	2	3	1		8	9		
	5	9	3			2	5	
	7	2		4	3	1		
3	1	8	5	7	9		7	9
1	2	5		6	2		6	5

#8

7	6	8		9	5	8	7	
9	3	6	5	8	2	4	1	
	2	1	4	6		8	4	
		7	9		3	4	5	1
6	9			4	8	9	2	
3	2	1		3	1	7		
	8	6	4	7	2	9	1	
		2	7	9		6	3	7
7	2	8	9				7	9
2	1	7	8		9	5		
9	3			1	8	3	9	
	5	9	7	4	6	1	8	3
	4	7	1	2		2	6	1

#9

			9	8				
8	5	3	7	1		4	8	
9	7	4	8			3	2	1
7	2	1		6	2		9	7
	3	2	4	9	1	7	5	8
1	4	5	6	8	3	9	7	
8	9		1	3		8	1	2
2	6	1		2	4	3	1	
8	2		8	1	5	6	3	
		9	8					

#10

9	8	5		2	7	1	3	
8	5	2	3	6	9	7	4	
1	5		1	2	4		9	7
2	6	8	3	5	1		8	9
	9	8	6					
			9	5	8			
1	8		7	8	1	4	9	3
2	5		9	7	6		4	1
6	7	3	8	1	2	4	5	
7	9	6	5		9	8	7	

#11

1	5			5	7	8	
3	9	4		2	1	6	8
	3	1	7	9		2	7	9
		1	3			4	6
9	5		6	8		8	5	1
3	2	8		7	9	4	2	
	1	4	9	5	8	2	3	
	3	1	5	2		5	1	2
1	4	9		4	6		9	7
2	6			1	7			
3	8	6		6	8	7	9	
5	9	8	6			3	5	8
	7	9	1			1	3	

#12

1	4		2	5	1	3		
3	5	1	4	9	2	8	6	7
		3	8		3	6	1	2
1	2	4	3		6	9		
5	7		7	2			5	9
	6	2	5	1		5	4	1
1	8	4	9	6	5	7	2	3
2	3	1		4	9	2	1	
4	5			3	8		9	8
		1	2		1	2	3	5
8	4	9	5		6	7		
4	2	6	3	9	7	1	5	8
		4	1	7	2		6	9

#13

		8	9	6			7	9
	9	6	7	2		7	9	8
8	4	1			2	6	4	1
9	7	3			6	9	8	
6	5		2	4	1	8	3	5
7	2	4	1	5	3		6	9
	8	9	7			9	2	6
5	1	3	4			2	1	3
9	3	8		7	9	8	5	
8	6			1	7	5		

#14

		8	7	9			5	3
		2	4	1		1	3	2
3	2	7	5			7	5	4
2	4	1		3	5	9	8	6
5	9		2	1	4		9	3
6	5		8	2	9		2	1
1	7	8	3	5		8	6	2
	8	9	6		8	9	7	4
4	6	7		1	7	3		
8	3			3	5	7		

#15

3	9	7		2	6		2	8
1	5	4		7	9		5	6
		3	7	1	8	6	4	5
	8	6	9	3		2	3	7
	4	1	2				7	9
			5	9	8	7	6	3
	9	3	6	8	7	2	1	
5	3	1	4	6	2			
9	7				1	3	7	
7	5	3		7	6	8	9	
8	6	1	7	5	3	9		
6	1		9	8		2	9	1
4	2		3	1		1	7	3

#16

3	9		3	8		1	9	
2	7	5	1	6	9	3	8	
	5	9		4	6	2	5	9
6	3	2	1	9	8		2	7
8	4	7	5		1	3		
9	6	8		6	2	4	9	3
3	2		7	8	4		5	2
7	1	2	9	4		2	3	1
	1	3		2	7	8	6	
4	2		8	3	5	9	7	4
6	3	7	4	1		8	6	
	1	9	5	2	3	6	4	7
	4	8		4	9		1	2

500++ Medium-Hard Kakuro by amazon.com/djape, page 134

#17 #18 #19 #20

#21 #22 #23 #24

#25 #26 #27 #28

#29 #30 #31 #32

#33 #34 #35 #36

#37 #38 #39 #40

#41 #42 #43 #44

#45 #46 #47 #48

#49 #50 #51 #52

#53 #54 #55 #56

#57 #58 #59 #60

#61 #62 #63 #64

#65　#66　#67　#68

#69　#70　#71　#72

#73　#74　#75　#76

#77　#78　#79　#80

#81 #82 #83 #84

#85 #86 #87 #88

#89 #90 #91 #92

#93 #94 #95 #96

#97　#98　#99　#100

#101　#102　#103　#104

#105　#106　#107　#108

#109　#110　#111　#112

#113　　#114　　#115　　#116

#117　　#118　　#119　　#120

#121　　#122　　#123　　#124

#125　　#126　　#127　　#128

#129 #130 #131 #132

#133 #134 #135 #136

#137 #138 #139 #140

#141 #142 #143 #144

#145 #146 #147 #148

#149 #150 #151 #152

#153 #154 #155 #156

#157 #158 #159 #160

#161 #162 #163 #164

#165 #166 #167 #168

#169 #170 #171 #172

#173 #174 #175 #176

#177

#178

#179

#180

#181

#182

#183

#184

#185

#186

#187

#188

#189

#190

#191

#192

#193 #194 #195 #196

#197 #198 #199 #200

#201 #202 #203 #204

#205 #206 #207 #208

#209 #210 #211 #212

#213 #214 #215 #216

#217 #218 #219 #220

#221 #222 #223 #224

#225

#226

#227

#228

#229

#230

#231

#232

#233

#234

#235

#236

#237

#238

#239

#240

#241 #242 #243 #244

#245 #246 #247 #248

#249 #250 #251 #252

#253 #254 #255 #256

#257 #258 #259 #260

#261 #262 #263 #264

#265 #266 #267 #268

#269 #270 #271 #272

#273 #274 #275 #276

#277 #278 #279 #280

#281 #282 #283 #284

#285 #286 #287 #288

#289

#290

#291

#292

#293

#294

#295

#296

#297

#298

#299

#300

#301

#302

#303

#304

#305   #306   #307   #308

#309   #310   #311   #312

#313   #314   #315   #316

#317   #318   #319   #320

#321 #322 #323 #324

#325 #326 #327 #328

#329 #330 #331 #332

#333 #334 #335 #336

#337 #338 #339 #340

#341 #342 #343 #344

#345 #346 #347 #348

#349 #350 #351 #352

#353 #354 #355 #356

#357 #358 #359 #360

#361 #362 #363 #364

#365 #366 #367 #368

#369 #370 #371 #372

#373 #374 #375 #376

#377 #378 #379 #380

#381 #382 #383 #384

#385 #386 #387 #388

#389 #390 #391 #392

#393 #394 #395 #396

#397 #398 #399 #400

#401 #402 #403 #404

#405 #406 #407 #408

#409 #410 #411 #412

#413 #414 #415 #416

#417 #418 #419 #420

#421 #422 #423 #424

#425 #426 #427 #428

#429 #430 #431 #432

#433 #434 #435 #436

#437 #438 #439 #440

#441 #442 #443 #444

#445 #446 #447 #448

#449 #450 #451 #452

#453 #454 #455 #456

#457 #458 #459 #460

#461 #462 #463 #464

#465   #466   #467   #468

#469   #470   #471   #472

#473   #474   #475   #476

#477   #478   #479   #480

#481 #482 #483 #484

#485 #486 #487 #488

#489 #490 #491 #492

#493 #494 #495 #496

#497   #498   #499   #500

#501   #502   #503   #504

#505   #506   #507   #508

#509   #510   #511   #512

Other Kakuro books, in different formats and sizes:

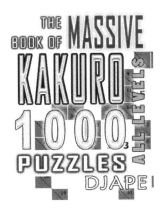

The Massive Book of Kakuro:
1000 puzzles of all difficulties
search this number on Google or Amazon:
1519348045

Difficult Kakuro:
search this number on Google or Amazon:
1493676873

Kakuro Puzzle Book:
search this number on Google or Amazon:
9798573381749

1,000++ Easy-Medium Kakuro:
search this number on Google or Amazon:
9798769433528

Mighty Kakuro 20x25 (**HUGE** puzzles!):
search this number on Google or Amazon:
1540579298

Made in the USA
Las Vegas, NV
10 December 2024

13756865R00092